T0223016

JCT Contract Administration Pocket Book

This book is quite simply about contract administration using the JCT contracts. The key features of the new and updated edition continue to be its brevity, readability and relevance to everyday practice. It provides a succinct guide written from the point of view of a construction practitioner, rather than a lawyer, to the traditional form of contract with bills of quantities SBC/Q2016, the design and build form DB2016 and the minor works form MWD2016. The book broadly follows the sequence of producing a building from the initial decision to build through to completion. Chapters cover:

- Procurement and tendering
- Payments, scheduling, progress and claims
- Contract termination and insolvency
- Indemnity and insurance
- Supply chain problems, defects and subcontracting issues
- Quality, dealing with disputes and adjudication
- How to administer contracts for BIM-compliant projects

JCT contracts are administered by a variety of professionals including project managers, architects, engineers, quantity surveyors and construction managers. It is individuals in these groups, whether experienced practitioner or student, who will benefit most from this clear, concise and highly relevant book.

Andy Atkinson is a chartered quantity surveyor with a background in consultancy and public service. With 35 years' experience lecturing in contract administration at London South Bank University, he has acted as principal and co-investigator for several publicly funded research projects. Formerly a member of the Joint Contracts Tribunal BIM Working Group, examining methods for adapting JCT contracts to building information modelling, Andy maintains a small surveying consultancy.

JCT Contract Administration Pocket Book

Second Edition

Dr Andy Atkinson
PhD, MSc, FRICS, Cert Ed.

Routledge
Taylor & Francis Group

LONDON AND NEW YORK

Second edition published 2021
by Routledge
2 Park Square, Milton Park, Abingdon, Oxon, OX14 4RN

and by Routledge
52 Vanderbilt Avenue, New York, NY 10017

Routledge is an imprint of the Taylor & Francis Group, an informa business

First edition published by Routledge 2015

British Library Cataloguing-in-Publication Data
A catalogue record for this book is available from the British Library

Library of Congress Cataloging-in-Publication Data
Names: Atkinson, Andy, author.
Title: JCT contract administration pocket book / Andy Atkinson.
Description: Second edition. | Abingdon, Oxon ; New York, NY : Routledge,
2021. | Includes bibliographical references and index.
Identifiers: LCCN 2020032542 (print) | LCCN 2020032543 (ebook) | ISBN
9780367638078 (hardback) | ISBN 9780367632786 (paperback) | ISBN
9781003120797 (ebook)
Subjects: LCSH: Building--Superintendence--Great Britain--Handbooks,
manuals, etc. | Construction industry--Subcontracting--Great
Britain--Handbooks, manuals, etc. | Construction contracts--Great
Britain--Handbooks, manuals, etc.
Classification: LCC TH438 .A87 2021 (print) | LCC TH438 (ebook) | DDC
692/.80941--dc23
LC record available at https://lccn.loc.gov/2020032542
LC ebook record available at https://lccn.loc.gov/2020032543

ISBN: 978-0-367-63807-8 (hbk)
ISBN: 978-0-367-63278-6 (pbk)
ISBN: 978-1-003-12079-7 (ebk)

Typeset in Goudy and Frutiger
by KnowledgeWorks Global Ltd.

Contents

Foreword vii

Preface ix

Introduction 1

1 **Building procurement strategy** 5

2 **Building procurement procedures** 37

3 **Interim payments** 59

4 **Final accounts** 73

5 **Progress** 87

6 **Claims** 97

7 **Termination and insolvency** 109

8 **The supply chain and subcontracting** 125

9 **Indemnity and insurance** 143

10 **Fluctuations** 161

11 **Maintaining quality** 169

12 **An introduction to dispute resolution in construction** 183

13 **The construction act – adjudication and payment** 195

14 **Building information modelling and the JCT contracts** 205

Glossary 221

Index 233

Foreword

The Joint Contracts Tribunal contracts remain the most widely used standard forms of building contract in the United Kingdom. About 80% of contracts are let using JCT forms with traditional forms predominating, closely followed by design and build. The forms now also provide the basis for many other standard contracts worldwide. The reasons for such wide use are based partly on their longevity, with a history stretching back over 80 years. This has the great advantage of allowing legal precedent to develop, ensuring that wording in the forms is "tried and tested" in the courts. It also has the operational advantage of familiarity for the professionals who are required to implement its provisions.

Two key characteristics of the JCT forms are, first, their collaborative nature at the institutional level. The JCT comprises representatives from all parts of the construction industry and clients including the British Property Federation, the Local Government Association, the RIBA, RICS, Contractors Legal Group and Build UK. This collaboration ensures that the forms are fair and accurately reflect the requirements of the industry. Second, the forms clearly allocate responsibility between participating project members. They also allow for collaborative working at the project level, although there is a clear distinction between those charged with design, building and paying for building work and this reinforces both legal and operational certainty.

Despite these positive characteristics, building contracts are only as good as the people operating them and it is essential that those entering the construction industry and building-related professions quickly get a good grasp on the administration of their contracts. This book enables students and new practitioners to do just that. Starting with a strategic view of procurement, covering many types of contract and tender, it concentrates on pre- and post-contract procedures with the three most widely used JCT forms – Standard Building Contract with Quantities (SBC/Q 2016), the traditional form with quantities, Design and Build Contract (DB 2016), the design and build form, and Minor Works Building Contract With Contractor's Design (MWD 2016), the minor works form.

This second edition of the JCT Contract Administration Pocket Book maintains the succinct format of the first edition but updates to the 2016 suite of JCT contracts. The 2016 contracts make amendments to reflect feedback from practitioners and incorporate such changes as the provision to name specialists for work sections, revisions to payment provisions and adjustments to reflect the Construction (Design and Management) Regulations 2015.

Dr Andy Atkinson has over 40 years' experience in the construction industry, initially as a quantity surveyor, then as a lecturer in contract administration and project management at London South Bank University. He has a reputation for succinct delivery and this book, whilst comprehensive, is readable. It covers complex subject matter briefly but clearly and at a level which is manageable for student or practitioner new to the subject. If this is you, or if you just require a refresher on JCT contract administration procedures as currently practiced, I recommend this book.

Neil Gower, Solicitor
Chief Executive, The Joint Contracts Tribunal
August 2020

Preface

Successfully managing JCT contracts is a must, and this handy reference is the swiftest way to doing just that. Making reference to best practice throughout, the book uses the JCT Standard Building Contracts SBC/Q, DB and MWD as examples to take you through all the essential contract administration tasks, including:

- procurement;
- payment;
- final accounts;
- progress, completion and delay;
- subcontracting;
- defects and quality control.

In addition to the day-to-day tasks, this book also gives an overview of what to expect from common sorts of dispute resolution under the JCT, as well as a look at how to administer contracts for BIM-compliant projects. This is an essential starting point for all students of construction contract administration, as well as practitioners needing a handy reference to working with JCT contracts.

This second edition updates practice to incorporate the latest 2016 editions of the JCT contracts, but without major changes. One significant update, however, is to include the Minor Works Building Contract, With Contractor's Design 2016 in place of the Minor Works Building Contract 2011. The former adds the facility to separately let sections of work on a design and build basis (As Contractor's Designed Portions). This is very useful in practice as, even for small projects, sections are let in reliance on the contractor carrying out design. Common sections include plumbing and heating, electrical installations, damp and rot treatment. Clearly identifying CDPs for this work ensures that design liability is addressed and adequately apportioned, so this form is to be preferred over MW2016 unless all work is to be directly designed by the Architect. Other revisions include updating Chapter 14 on BIM and the JCT contracts to refer to the second edition of the CIC BIM protocol and reference to the Public Contracts Regulations in Chapter 7 on Termination.

Introduction

Commercial contracts are how buildings are built! No matter how good the design, engineering and construction, without arranging a suitable means for specifying, ordering and paying for work, nothing will happen. This means that it is vitally important to understand how to arrange and administer building contracts. Without this understanding, whether you are a client, architect, quantity surveyor, contractor or sub-contractor, you might not get what you expected. Additionally, the fact that contracts involve different parties with different interests leads to a natural tension between client and contractor. A client will naturally want the best building as quickly as possible at the lowest possible price. If a client fails to do this, it will be accused of not obtaining value for money. A contractor must be commercially oriented towards maximising profit for the firm. If a contractor fails to do this, its share price will dip and it will be vulnerable to take over from another firm, perhaps with a more robust intent! Although it is in the long-term interest of all parties to produce good buildings, it is naïve to expect all parties in a project to work to the benefit of the project and deny that strong commercial pressures exist to extract the most from the other party. Cooperative forms of working such as partnering based on negotiation are, therefore, of limited practical applicability. Sooner or later, one party will feel aggrieved that they are being taken advantage of by the other and the love affair will be over. This falling out is often seen at the start of a recession in the industry, when margins tighten and pencils are sharpened. Parties are happy to cooperate when there is plenty of work to go around, but as soon as profits or cash flow are threatened, they start to look for any additional benefits by scouring the contract.

Once the natural tension in building contracts is recognised, it is easier to understand how performance can be improved through contract administration. Clear divisions of responsibility and good preparation are particularly important. "Fast-tracking" – the process by which design, logistics and construction are overlapped in order to save time – was criticised over 25 years ago for just this reason[1]. It allows construction to start before the design is properly finalised and work is fully quantified. Getting the design properly sorted out, whether by a consultant architect/engineer team or by a contractor's design team, gives clarity to the constructors, leaves less to be muddled through on site and generally makes a project run better. Technically well-produced and fully specified designs based on properly researched sites, produced by competent designers and constructed by skilled craftspeople will make good buildings. No contract will change that, but some can encourage it, by setting out clearly the expectations of the parties, the course of action to be taken if things go wrong and the remedies to be available if one party does not do as agreed. The Joint Contracts Tribunal (JCT) standard forms of building contract seek to do this.

In any argument about what was agreed, the detail of the agreement is likely to be very closely scrutinised, with increasing diligence as the argument deepens. If the argument ends up in court, deciding what was actually meant by a particular turn of phrase in the contract will be based, where possible, on precedent. Although the wording of contracts is often arcane and reads peculiarly in a modern context, it may be left unamended for this reason. The courts, lawyers and practitioners know, based on precedent, the meaning of terms. This is another of the key advantages of the JCT contracts and their antecedent (the Royal Institute of British Architects standard form of contract). These were first produced in the 19th century and, although described by one judge as a "farrago of obscurities"[2], they contain tried-and-tested obscurities, familiar to both practitioners and lawyers, if not to high court judges. JCT contracts have now been thoroughly modernised and wording has been brought up-to-date, but without losing the basic familiarity of earlier forms to the extent that many procedures in the industry have the feel of being custom and practice although actually based on contract requirements.

This book is about contract administration using the JCT contracts. It provides a succinct guide written from the point of view of a construction practitioner, rather than a lawyer, and is not intended to provide a legal guide to the forms. Nor does it cover all detailed aspects of administration, excluding particularly the detailed requirements of the Construction Design and Management Regulations[3], which are better dealt with in specialist texts. The JCT now produces a very wide range of standard forms of contract and coverage of the book is limited to the three most widely used – the traditional form of contract with bills of quantities SBC/Q2016, the design and build form DB2016 and the minor works form (with Contractor Design) MWD2016. These three forms of contract are used in well over half of building and engineering projects in the United Kingdom[4] and form the basis for many other standard contracts worldwide. As with most sets of standard forms of contract, they have strong family likenesses and all originally stemmed from the traditional form. They are administered by a variety of professionals, including architects, building surveyors, engineers, quantity surveyors and construction managers. It is towards individuals in these groups, whether experienced practitioner or student, that the book is directed.

The book broadly follows the sequence of producing a building from the initial decision to build to completion. It starts by considering procurement strategy where the objectives of the client are matched to available procurement choices. Procurement procedures are then reviewed from the decision to invite tenders to placing the building contract.

Once contracts are entered, money and time are themes close to the heart of both clients and contractors, so interim and final payment, progress, extensions of time, fluctuations and claims are dealt with in detail. Occasionally, the employment of a contractor is terminated (often precipitated by the insolvency of one of the parties) and this is also covered.

Buildings are exceptional in that they are usually not occupied by the customer. The building is passed on to a further purchaser or tenant, who also may not be in occupation. This gives rise to supply chain problems, particularly in relation to dealing with latent defects in the building that are the liability of the original builder or sub-contractor. Dealing with these problems and other subcontracting issues form further chapters of the book.

Defects are also an issue when they occur during the progress of the work. The client should not have to pay for them and they should be put right, but sometimes they are not discovered until later and they are not particularly problematic. Should they be put right or could a client accept non-conforming work at a discount? Tests are often pre-ordered or requested on an unplanned basis, but work fails. Questions arise as to who pays should subsequent tests – perhaps on the finished structure – pass. These considerations form a further separate chapter.

Another important area of practice relates to insurance. Clearly, it is important to insure public liabilities for injury to third parties passing a building site and for damage to third party property, but why this insurance should be a matter for a building contract is less clear. Similarly, a contractor would normally have to complete a building even though it is damaged by an intervening event such as a fire. But works insurance is a matter for a building contract and of interest to the client and consultants as well as the builder. It is possible for design and other consultants to insure for negligence in performing their duties by taking out professional indemnity insurance and understanding, in outline, how this insurance works is important to all construction professionals. These and other insurance-related matters are contained in the chapter on insurance.

Although building contracts normally go as planned, occasionally serious disagreements arise. The complexity of construction has meant that special procedures have been developed when this happens. It is now common to try and settle disputes amicably, either directly or with assistance from a mediator or conciliator. If that is not possible, reference to a formal adjudicator may be required and should one or other party not like the adjudicator's decision, they may appeal to arbitration or to a court. Dealing with disputes is contained in a separate chapter, with a further chapter covering construction industry adjudication in more detail.

Finally, conceptualisation and documentation in the construction industry is being revolutionised by the introduction of powerful modelling software. The advent of Building Information Modelling (BIM) has introduced a whole new vocabulary and, although not necessarily changing radically the specialisms within the industry, it will change the way information is presented in contracts. Gone will be contract documents – drawings, specifications, bills of quantities, schedules of rates – replaced by federated and specialist models. The printed output (if any) may take the same form as existing documents, but their production and the underlying capabilities of the models producing them will be radically different. The impact of BIM on the three JCT contracts, as represented by a widely adopted BIM protocol, is reviewed in a separate chapter.

NOTES

1. Tighe J J (1991) Benefits of Fast Tracking are a Myth, International Journal of Project Management, Vol. 9 Issue 1, 49–51.
2. Per Davies LJ in English Industrial Estates Corporation v George Wimpey and Co Ltd, 1972, 7 BLR 122 at 126.
3. UK Government (2015) The Construction (Design and Management) Regulations 2015, SI 2015 No. 51, The Stationery Office, London UK, at http://www.legislation.gov.uk.
4. NBS (2018) National construction contracts and law report 2018, RIBA Enterprises Ltd, London UK, at https://www.thenbs.com.

1

Building procurement strategy

INTRODUCTION

What is procurement?

In life, important decisions are made early. Careers and income are usually determined by choices made at school and university and often these choices are based on minimal information. Similarly, with building projects, the big decisions come at the beginning, not least the decision to build at all. Once the decision to build is made, it rules out all other options, including the sometimes very attractive options to do nothing, to buy a ready completed building, or to invest in reorganisation.

These very early decisions are more the province of executive project management than contract administration, but the decision to build still leaves many decisions on more detailed choices – how to build, who will do the work and how will the participants in building be organised. There were not always choices and in many countries until recently there was really only one way of getting buildings built. With no choices, there were no problems to solve and little consideration was given to how projects were organised. Now that there are a number of ways with which the key players in building projects can be matched up and key tasks can be sequenced, procurement decisions have to be made. Procurement, therefore, involves making big and influential early decisions, which determine the overall pattern of building – how the work is to be done and who is to do it. Not the biggest decision of whether to build at all, but nonetheless, important in the context of getting the building expected.

Who are the key players?

One thing which is common to any choice of procurement is that the basic ingredients of the building process do not alter that much. It is necessary to plan, design and construct the building. The prime method of doing this is through commercial contracts between key players. The players may work individually but more often are employees of commercial organisations and public bodies. The key players include the following.

Clients

The client is the initial owner and commissions the building work, but may not be the driving force behind the project, or the eventual owner. In building contracts, the client is often termed the Employer (UK) or the Owner (USA). The initial building owner often never occupies the building and may be a developer, who deals in land and buildings. Other key players on the client's side include the eventual owner of the building, who could be an investment company, the first tenant and subsequent tenants and the user. Depending on the type of building, these players could all be separate, and sometimes very large, organisations.

Designers

Designers conceptualise the building in a form that can be understood by both client and builder. Sometimes during the building work they also supervise work to ensure that it is built as intended. The key designer in building projects is often an architect, but (for refurbishment and repair projects) may be a building surveyor. In civil engineering projects, the key designer is normally a civil engineer. In all but the smallest building project, the key designer is supported by other specialists. These include structural engineers (designing structural elements of the project), services engineers (designing the building services such as air conditioning) and landscape architects (designing surrounding grounds).

Quantity surveyors

A quantity surveyor, or taker-off, quantifies the work shown on the drawings in a way that it can be priced by the builder. Quantity surveyors also provide estimates of the likely cost of building work before detailed design and may supervise payment of amounts to builders on account. The quantified work may (depending on the procurement choice made) be presented in "bills of quantities" to separate builders to price, or as internal documents prepared by the builders ("builders quantities").

Builders

The builder is responsible for building in accordance with the design. Builders are often termed "contractors" because the way they are employed is by a commercial contract. Modern building companies often concentrate on managing building operations and leave most actual construction to separate firms of "specialists". The specialists concentrate on sections of the work – for example on the sub-structure work below ground level, or the concrete frame. Key operations within the building company will include estimating the cost of the work, construction management (sequencing and controlling operations) and, for some procurement choices, taking off quantities for the estimators. Some firms may also undertake design work.

Specialists

Specialist building firms carry out sections of the work. These firms usually enter into a contract with the builder and are termed "sub-contractors", but they may also, depending on the procurement choice made, contract directly with the client, or have two contracts, one with each. Some specialists for complex sections may also sub-contract work further to sub-specialists.

Authorities

Local authorities provide the regulatory context for carrying out building work. In particular, they are responsible for granting planning permission and building regulations approval. They also have some control over the nuisance that building operations can cause. Planning permission is a political decision concerned with whether the building should be built at all and what form it should take. Getting planning permission can be a lengthy process (in some instances taking several years) and can require considerable details of the design of the building. Work cannot start until permission is obtained. Building regulations approval is concerned with the technical detail of the building and is also often obtained before work starts. Building regulations approval is not a political decision and, in some instances, can be given during progress of the work, so it is not quite so restrictive to building operations as planning permission.

What procurement choices are there?

The way a project manager might approach deciding on a procurement path would be to look at the objectives of the client and match the procurement choice to the objectives. The problem is that the objectives are all likely to go in the same direction – the best building possible, as quickly as possible, at least cost and with no financial or other risk (objectives are usually summarised as time, cost, quality and risk). Different procurement choices emphasise these objectives differently. Thus, one choice might make it easier to ensure the best building possible, but at greater cost, time and financial risk than another. Although it is better to work from objectives to procurement choice in this way, for a newcomer to the subject it is easier to invert the problem and take a solutions based approach. It is better to look at the choices available and then consider their advantages and disadvantages. Looking at solutions is more familiar to contract administrators, who live with the detail of different procurement choices, and so could be termed a "contract administration approach".

Using this approach for building projects, "how work is to be done" translates into the <u>choice of building contract</u> and "who is to do the work" translates into the <u>choice of tender type</u>.

BUILDING CONTRACTS

Lump-sum or cost reimbursement?

Building contracts can be categorised in a number of ways including whether they are written for a particular project, or are a standard form written for use on several projects. However, the legal "form of contract" is not of interest here. Rather, it is the overall type of contract that is of interest. That is, within whatever form is being considered, how the key players are organised and their work sequenced. The first broad consideration is whether the contract is a lump-sum or cost-reimbursement type. Cost-reimbursement contracts have a long history going back to the construction of medieval cathedrals and before. They consist broadly of agreements to pay for the constituents of building (labour, material and plant) at fair prices as they accrue. This method of contracting served well up until the industrial revolution but demands to invest in buildings as a means of producing the artefacts of a machine age meant more certainty of cost was needed. In response to these demands, builders began to offer to build to a price. They were prepared to take a view on the risks and costs involved and the profession of estimating was born. Buildings were not necessarily any more expensive, but builders were undertaking to shoulder the financial risk of buildings being more difficult to build than expected when the cost was estimated. Most contracts (including the three reviewed in detail in this book) are now of this lump-sum type, but cost-reimbursement contracts still have a place where it is not possible or is unreasonable to ask builders to provide lump-sum estimates for carrying out work. Additionally, hybrid contracts are now quite widely used. These have elements of both cost-reimbursement and lump-sum agreements. A modern hybrid contract type is management contracting, where the management contractor is paid on a basis broadly similar to cost reimbursement, but the specialist contractors actually carrying out the work are on lump-sum contracts.

Lump-sum contracts

Within this broad category, types of contract can be further divided according to how much financial risk is transferred from the building client to the contractor. With those that transfer most risk at the top, a list of lump-sum contracts would involve:

1. PFI (private finance initiative) public sector contracts
2. Design and build contracts
3. Traditional drawing and specification (without quantities) contracts
4. Traditional contracts with quantities
5. Measured-term contracts

PFI public sector contracts

WHAT ARE THEY?

These contracts involve a contractor providing the building design, construction, maintenance, facilities management and, sometimes, staffing in exchange for payment in the form of rent over a defined period (often 30 years). At the end of this period, the public sector client retains the building. The contractor is, therefore, required to take not only the risk of construction costs being to the estimate but also maintenance, facilities management and staffing costs. As these contracts involve at least three separate specialisms of building, finance and facilities management, it is common for larger projects to be joint ventures using companies floated, especially for the purpose (special project vehicles). Shares (stock) in the vehicle are held by all major specialists. The construction will then be sub-contracted to the building company. The client is normally a public sector body such as a health service organisation, a public airport operator or a local authority and can, therefore, guarantee payment of the rent.

A PFI contract will be specified by the client in terms of the functional output required (e.g. the number of places for a school building), but, as the client will retain the building at the end of the defined period, details of the design and specification proposed by tendering contractors will also be important. Letting PFI projects is therefore complex, involving consideration of rent offered, design and the cost of running and staffing the building.

PFI contracts are on the boundary of what would be considered building contracts. They involve payment of rent in exchange for use of a building over a long period and thus are similar to commercial property transactions. However, they are direct alternatives to simpler arrangements where building owners contract with builders. It is for this reason that they are included here.

PFI CONTRACTS – KEY DECISION FACTORS

Advantages	1 Transfers construction and operating risks from client to contractor 2 Does not require capital expenditure from the client 3 Promotes efficient building by centralising design, construction and operation with the contractor 4 Promotes efficient buildings by encouraging the bidding contractors to consider whole life costs (cost of building, running, maintaining and demolishing the facility)
Disadvantages	1 Inflexible once project is let and cannot easily handle changed circumstances over the lease period – both parties may be tied into a long lease 2 Finance can be more expensive as contractor usually will not get cheaper financing rates than a public client 3 Indexed rental payments may increase more rapidly than client financing rates making the project more expensive for the client
Suitable circumstances	1 Any public projects with absolute limitations on capital expenditure 2 Projects with clear and consistent operational requirements
Most suitable projects	Prisons, hospitals, roads and railways

Design and build contracts

WHAT ARE THEY?

In these contracts, a builder executes both the design and construction of a building to a client's brief. There is no definition of what the brief contains and it could be a simple verbal or written statement of requirements in performance terms (a performance specification), or it could involve more detail.

PURE DESIGN AND BUILD

If the brief is expressed solely in performance terms, the builder would be involved in producing both outline designs and detailed designs for construction. The client would be relying on the skill of the builder, both in designing and building, and could expect the builder to produce a building which is fit for the purpose expressed in the brief. This type of contract, sometimes termed "pure design and build", centralises liability for the building with one party – the builder. This minimises complications should anything go wrong after construction. Although many design and build contractors have in-house designers, often design is subcontracted to consultant architects and engineers. A common arrangement is to sublet the outline design, but to carry out detailed design in-house.

SCOPE DESIGN AND BUILD

Clearly, a client who lets a project on the basis of a performance specification will have little direct control over the design. Whilst this may be adequate for many commercial or industrial buildings, many clients will wish to work out the design in advance directly with an architect. Thus, many design and build projects include outline or conceptual design as part of the brief, with the builder being responsible for carrying out detailed design and construction. This is a popular arrangement where statutory approvals such as planning permission involve detailed negotiation with public authorities before the building contract can be let. It is also popular with developer clients who want to work out designs in order to maximise the sale or letting potential of the finished building but are less interested in detailed design. The outline design will typically provide sufficient information to satisfy planning authorities of the form and nature of proposals and will give enough information to attract prospective purchasers or tenants. Where outline designs are carried out by the client's consultant architect, liability for the building may be split between architect and contractor and the advantage of centralised liability is lost. For work designed by the consultant, it may be difficult to pinpoint whether a failure is due to poor design, poor construction, or even poor supervision by the architect. To avoid such splits of liability, many clients, particularly developers, re-centralise liability by using a "novation" device, which requires the appointed contractor to take on as a sub-contractor the consultant architect. Novation is considered in more detail in Chapter 2.

DESIGN AND BUILD CONTRACTS – KEY DECISION FACTORS

Advantages	1	Certainty of time, cost and quality engendered by inflexible contractual arrangements
	2	Only one party to deal with providing administrative simplicity and centralised liability
	3	Stronger legal liability when builder is required to produce a building fit for its purpose
	4	Efficient design/build communications promotes lower prices and quicker building

	5 Design and construction operations can often be overlapped, where work is priced on outline information, promoting quicker building 6 Promotes innovation in design leading to cheaper and quicker building
Disadvantages	1 Harder to select a builder as tenders are based on two factors – design and cost 2 Tendering is expensive as every contractor has to design, quantify and price the project 3 Inflexible as the client has less tools to economically vary the works 4 Designs may be uninspiring, emphasising the efficiency of construction at the expense of design quality 5 Still requires consultancy input to assist the client with contractor selection and supervision 6 Lack of redress to correct latent defects should builder become insolvent
Suitable circumstances	1 Simple buildings with clear design requirements, which are unlikely to change 2 Projects where the client wishes to transfer design and construction liabilities to a single party 3 Projects where time and cost savings are primary objectives
Most suitable projects	Commercial and industrial buildings intended to produce revenue – warehouses, factories and simple office buildings

Traditional drawing and specification (without quantities) contracts

WHAT ARE THEY?

The traditional pattern of contracting in many parts of the world, but especially in Europe, is to engage consultant designers to carry out both conceptual and detailed design and to separately engage building contractors to carry out the building work itself. This pattern takes many forms for building and civil engineering projects, both related to lump-sum and cost-reimbursement contracts, but a very common arrangement is to provide design in the form of detailed drawings and a separate written specification. The builder is required to offer a lump-sum price for carrying out the work and therefore takes most of the risks of building including risks associated with accurately determining the quantity and cost of the work involved. However, (as between builder and client) risks related to design are retained by the client.

A feature of traditional contracting is that, in addition to designing the work, the designer will carry out some post-contract supervision to ensure that the building is constructed as designed and authorise interim and final payment for the building work. Liability for the performance of the building is split between designer and builder and, as with scope design and build, it is sometimes difficult to pinpoint who has been responsible for a failure – the designer (in initial design or supervision) or the builder.

Traditional drawing and specification contracts – key decision factors

Advantages	1 Independent consultant design advice backed by professional indemnity insurance 2 A fully specified and detailed design before going out to tender 3 Cost efficient design and specification – only one design is produced 4 Easy to select a builder as tenderers are not pricing for design 5 Consultant designers are more interested in the effectiveness of the building than the efficiency of construction and this may improve quality
Disadvantages	1 Liabilities are split between designer and builder leading to less effective redress should things go wrong 2 Weaker legal liability for both designer and builder – neither can be required to produce a building that is fit for purpose 3 Communications between separate designer and builder are bureaucratic and cumbersome 4 Multiple parties for the client to deal with 5 Less certain in terms of cost and time, where design is varied after work starts 6 Design and specification need to be fully developed before inviting tenders and starting work. This prolongs the overall duration of the project
Suitable circumstances	1 Projects where the client requires control over the detail design and specification of the project 2 Smaller projects not needing the flexibility induced by using bills of quantities (see below)

Most suitable projects	Any smaller building work with high design content including technically complex repair and refurbishment work requiring specialist design services

Traditional contracts with quantities

WHAT ARE THEY?

One of the problems with letting projects using design and build or traditional contracts without quantities is that each builder has to calculate the quantities of work required before providing an estimate. Taking off quantities from drawings is highly skilled, error prone and costly, so the process is sometimes carried out by a specialist consultant quantity surveyor and the quantities are provided to all contractors for pricing. The client guarantees that the quantities are correct, or if there are errors, these will be corrected. The builders are relieved of the cost and risk associated with calculating quantities and, therefore, can price more keenly. Although the client effectively takes the risk of errors in quantities, it also gets, when the builders return priced bills of quantities, detailed pricing information. This is useful for interim certificates paying for work on account, valuing variations to the work and for estimating and cost-planning future projects.

A variant of traditional contracts with quantities is to use "approximate quantities". Bills of approximate quantities will be produced in much the same way as above, but the items and quantities will be estimated, usually on the basis of past similar work combined with preliminary designs. Detailed designs are completed as the building work is carried out and the approximate quantities are firmed up by re-measurement. This overlapping of design and construction allows a considerable saving in time.

TRADITIONAL CONTRACTS WITH QUANTITIES – KEY DECISION FACTORS

Advantages	
	1 Independent consultant design advice backed by professional indemnity insurance
	2 A fully specified and detailed design
	3 Cost efficient design and specification – only one design and one set of quantities are produced
	4 Easy to select a builder as tenderers are not pricing for design
	5 Consultant designers are more interested in the effectiveness of the building than the efficiency of construction and this may improve quality
	6 Flexible post-contract as bills of quantities provide a detailed schedule for valuing variations

	7 Bills of quantities allow accurate assessment of work done for interim payment purposes 8 Bills of quantities provide pricing data for estimating and cost planning future projects
Disadvantages	1 Liabilities are split between designer and builder leading to less effective redress should things go wrong 2 Weaker legal liability for both designer and builder – neither can be required to produce a building that is fit for purpose 3 Communications between separate designer and builder are bureaucratic and cumbersome 4 Multiple parties for the client to deal with 5 Much less certain in terms of cost and time, where design is varied 6 Design and specification need to be fully developed and bills of quantities prepared before inviting tenders and starting work. This prolongs the overall duration of the project
Suitable circumstances	1 Projects where the client requires control over the detail design and specification of the project 2 This type of contract is suitable where the brief is likely to be evolving and substantial changes during construction may be necessary
Most suitable projects	Landmark buildings by signature architects, prototype buildings where changes are needed to allow for evolving configuration and buildings where requirements for use may change during the construction period

Measured-term contracts

WHAT ARE THEY?

This form of contracting also involves traditional arrangements of separate designer and builder. A measured-term contract, however, does not start with a defined scope in the form of drawings, specification or bills of quantities. It may only be envisaged that a certain level of work (normally small works or maintenance) will be required at a location or several locations over a period of time (the term). A solution to getting the work done, without having to let each item of maintenance/small works separately, is to ask builders to price expected items of work and, if the work arises within the term, to pay on the basis of the prices provided.

The way-term contracts are actually handled is to provide each builder pricing for term contracts with a schedule of rates. This document contains all the items of work ever likely to be needed for a project. The builders provide a price (rate) for each item (e.g. for supplying and laying a cubic metre of concrete), but there are no quantities of work associated with the items (no work has as yet arisen!). A consultant to the client (usually a quantity surveyor) will multiply the price of each item by "notional quantities" (likely quantities expected) to give a notional (theoretical) price for the whole project. The prices of each tendering contractor can then be compared and a suitable firm selected. The selected builder will then be required to carry out any or all work at the location for the rates it quoted for the duration of the term. This type of contract is popular with clients who have large estates, such as social housing providers or defence establishments. As with traditional contracts with quantities, the risk of the quantities being incorrect is borne by the client, but the builder is required to take the risk that the individual rates provided can be executed profitably. As the nature and quantities of actual work ordered are indeterminate, the client will take the risk that an overall programme of work can be carried out within the defined period.

MEASURED-TERM CONTRACTS – KEY DECISION FACTORS

Advantages	1	Allows the client to engage a single builder for a programme of work of several small projects
	2	Builders tender on the basis of rates, which they have to stand by, thus the client transfers pricing risk. Each individual rate is, effectively, a lump sum
	3	Very flexible as scope of work is not guaranteed
	4	Design need not be developed at the time of tendering and can be overlapped with construction, thus saving time
Disadvantages	1	Requires the production of schedules of rates
	2	Requires administrative machinery to organise tenders, place contracts and supervise work
Suitable circumstances	1	Large programmes of projects of a minor or maintenance nature <u>and</u>
	2	Defined types or location of work – for example, of a single construction type, or on a single site

Most suitable projects	Repairs, alterations and maintenance work to housing estates and other large estates such as defence establishments and industrial complexes

Cost-reimbursement contracts

Within this broad category, types of contract are further divided according to how complex the arrangements are. The simplest form of cost-reimbursement contract is for the client to pay for the cost of labour, materials and plant as it arises, with an addition for the builder's general overheads and profit. More complex arrangements introduce targets and yet more complex arrangements are hybrid types where elements of cost-reimbursement contracting are combined with lump-sum sub-contracting. The types considered below, therefore, are:

1. Simple cost-reimbursement contracts
2. Target cost-reimbursement contracts
3. Management contracts
4. Construction management contracts

Simple cost-reimbursement contracts

WHAT ARE THEY?

The arrangement of key players for this type of contract is normally the same as for traditional lump-sum contracts, but the basis of contracting is different. The builder is not required to provide an estimate for the cost of carrying out work. This cost is reimbursed (hence the term cost reimbursement) as it is accrued. Labour is paid, either at an hourly rate provided by the builder, or at nationally agreed rates. Materials and plant are paid on the basis of agreed invoices. Site overheads (those associated with the building site generally – such as site cabins, scaffolding and cranes) are also reimbursed in the same way. However, general overheads associated with the overall running of the building firm (such as the head office salaries and expenses), and profit are added to the cost as a percentage.

The percentage addition is the only competitive element in the builder's quotation as all other costs are simply reimbursed. However, the addition is not a good guide to the efficiency of the firm. A firm may charge a higher percentage addition for doing work but may be considerably more efficient in using resources than a rival with a lower percentage addition. The overall costs may therefore be lower with the apparently less competitive firm. In addition to this difficulty in selecting the builder, cost-reimbursement contracting directly rewards profligacy in that the higher the costs, the greater the percentage addition. In cost-reimbursement contracting, the client takes most of the time and cost risks of construction, including the efficient ordering and use of materials, plant and labour. In exchange, the builder is not required to price for these risks and this can produce cheaper building. Conversely, the client is not required to offer any particular scope of work, and if money runs out, the contract can easily be terminated.

SIMPLE COST-REIMBURSEMENT CONTRACTS – KEY DECISION FACTORS

Advantages	1	Selection of builder is based on overheads and profit only so can be made early
	2	Overlap of design and construction is possible, further saving time
	3	Flexible form of contract, with no guaranteed scope
Disadvantages	1	Difficult to select a suitable builder as only the level of overhead and profit is tendered
	2	Puts most risks of performance to time and cost on the client
	3	Provides an incentive for the builder to be profligate
Suitable circumstances	1	In emergency situations
	2	For investigative work, e.g. where the extent of repair has yet to be determined
Most suitable projects		Emergency repairs and opening up buildings to carry out further investigations

Target cost-reimbursement contracts

In order to try and combat the inefficiency of cost-reimbursement contracting target cost-reimbursement contracts have been developed. The principle of this type of contract is simple. The builder is rewarded with an enhanced addition for general overheads if costs are lower than expected. Two variants are commonly used, fixed-fee and true target cost contracts.

FIXED FEE

In this variant the addition for general overheads and profit is expressed and paid as a lump sum rather than a percentage. Thus, if the final cost of the project is less than expected, the fee increases in percentage terms. If the final cost is more than expected, the fee decreases in percentage terms. The introduction of "expected cost" introduces both a requirement for the client to estimate it and to stick by it.

Fixed-fee contracts are not, therefore, suitable for wholly indeterminate projects – there must be some realistic possibility of providing an estimated prime cost of the work. This will require expert evaluation, usually by a quantity surveyor carrying out an approximate estimate. Further, if the client fails to stick by the expected cost, for example by requiring additional work, the comparison of final cost with expected cost becomes invalid. The percentage payment to the builder will go down, not as a result of inefficiency, but because the client has increased the scope of the work. A fair comparison requires that client generated extras are removed from actual cost, giving a "notional cost" of what the work would have cost, but for the increased scope. This can then be compared with the estimated prime cost.

TARGET COST

True target cost contracts follow the same principle as fixed-fee contracts but gear up the fee. Should notional costs be lower than the estimated prime cost, the builder is rewarded with a share of the savings in addition to the fixed fee. For example, if notional costs are over the estimated prime cost, the contractor gets only the fixed fee. If notional costs are below the estimated prime cost, the contractor gets the fixed fee plus, say, 20% of the savings. This gives a stronger incentive to efficiency, but also means that the accounts will be much more carefully scrutinised by both client and builder.

TARGET COST-REIMBURSEMENT CONTRACTS – KEY DECISION FACTORS

Advantages	1 Selection of builder is based on overheads and profit only so can be made early 2 Overlap of design and construction is possible, further saving time 3 Flexible form of contract, with a limited guarantee of scope 4 Provides an incentive, in the target, for the builder to save cost
Disadvantages	1 Difficult to select a suitable builder as only the level of overhead and profit is tendered 2 Puts most risks of performance to time and cost on the client 3 An estimated prime cost must be calculated, thus the extent of work must be known 4 Notional and actual final accounts must be produced to be able to calculate overhead payments. This is administratively cumbersome

Suitable circumstances	1 Where it is necessary to start as soon as possible <u>and</u> 2 Where work can be defined in sufficient outline to determine an estimated prime cost
Most suitable projects	Refurbishment of larger commercial buildings where there is a known large revenue stream available as soon as the building can be occupied, which will offset any increase in capital cost associated with the system

Management contracts

Arrangements for simple management contracts can be very similar to traditional contracting, with separate consultant designers, or can be "design, manage and construct" contracts, where the builder also provides design services. This arrangement can be further complicated if the builder offers a guaranteed maximum price for the work.

SIMPLE MANAGEMENT CONTRACTS

In a simple management contract, the builder contracts with the client for a fee (to cover general overheads and profit) in much the same way as with cost-reimbursement contracts. The fee may be fixed or for a percentage. All work, other than that related to general overheads, is, however, sub-contracted to specialists. No building work (even that related to site overhead items) is carried out in-house by the builder. The client reimburses the costs related to the specialists to the management contractor. Specialist work is usually let as lump-sum contracts, but, in contrast to a normal lump-sum contract, the builder does not take the risk of the performance of the specialists. Provided the management contractor is not negligent in supervising the work, if specialist work is more expensive, is delayed or defective, the consequences are simply passed on to the client. Specialist firms in management contracts are called works contractors and, as they have no direct contract with the client and the contractor has no direct liability for their performance, there could be a problem if their work goes wrong at a later date. As a consequence, it is necessary to set up a direct contract (a collateral warranty[1]) between works contractor and client. Simple management contracting uses consultant designers and quantity surveyors in much the same way as with traditional contracts with quantities, but the role of the consultant quantity surveyor is more prominent. This is because the quantity surveyor will oversee the placing and financial management of all the works contracts as well as the management contract.

DESIGN, MANAGE AND CONSTRUCT

This type of contract is similar to design and build, but with the builder engaged on a management rather than a lump-sum contract. The builder is engaged for a fee to both manage and design the project and sublets all building work to specialists. A consultant quantity surveyor or similar financial expert is still needed in order to oversee placing and financial management of works contracts. Many clients will also appoint a separate consultant architect to provide scope designs or oversee the builder's design.

GUARANTEED MAXIMUM PRICE

Where the management contractor also has control of the design of the works it is possible for it to "design to a price". That is, the builder can develop and alter the design to ensure that the overall cost comes within a predetermined figure. This manipulation of the design is not possible where the builder does not control the design (e.g. where the designer is employed directly by the client). The predetermined figure (guaranteed maximum price) is usually set some way above expected final cost, but the builder agrees to stand by it and absorb the extra if costs turn out to be higher.

MANAGEMENT CONTRACTS – KEY DECISION FACTORS

Advantages	1 Selection of builder is based on overheads and profit only so can be made early
	2 Overlap of design and construction is possible, further saving time
	3 Offers some flexibility as works contracts can be let in stages
	4 Also offers some certainty in terms of performance as works contracts are usually let as lump-sum contracts
Disadvantages	1 Difficult to select a suitable management contractor as only the level of general overhead and profit is tendered
	2 Puts risk of performance of works contractors largely on the employer
	3 Provides little incentive to builder to be cost efficient as payment is fee based. There may be a tendency to use more expensive works contractors

	4 An estimated prime cost must be calculated as a basis for the management fee, which introduces rigidity
	5 Management contracting is opaque between employer and works contractor as far as financial arrangements are concerned (management contractor acts as a transfer payment service). This can lead to the use of hidden discounts and accusations of fraud
Suitable circumstances	1 Large indeterminate projects
	2 Complicated projects
	3 Major alteration projects
Most suitable projects	Large revenue earning commercial and industrial buildings, where time saved by the approach outweighs any likely cost disadvantages. Major complex refurbishment projects

Construction management

Construction management contracts dispense with a builder entirely and, instead, the client engages directly with specialists. However, there is still a need to plan, coordinate and control the building work. These functions are taken by a consultant construction manager, who has a similar contract to the design consultants. This type of contract can be very complex for the client as it will need to hold separate contracts for every specialist. The design team, including the construction manager and quantity surveyor, will advise on appointments for each specialist contract and provide overall supervision during the project. One advantage with construction management is the direct link between client and specialist. Removing the intermediate main or management contractor leads to direct legal liability, more transparency in financial dealings and, in many cases, faster payment to the specialist.

CONSTRUCTION MANAGEMENT – KEY DECISION FACTORS

Advantages	1 Selection of construction manager is based on managerial fee only so can be made early
	2 Overlap of design and construction is possible, further saving time
	3 Offers flexibility as specialists contracts are let in stages

	4	Also offers some certainty in terms of performance as specialists contracts are usually let as lump-sum contracts
	5	Contracts are direct between client and specialists aiding transparency
	6	Direct contracts between client and specialists are legally simple
Disadvantages	1	Difficult to select a suitable construction manager as selection is based on managerial skill (a less tangible quality) and not on the cost of building work
	2	Puts risk of performance of specialists largely on the employer
	3	Provides little incentive to construction manager to be cost efficient as payment is fee based. There may be a tendency to use more expensive specialists
	4	Administratively complex for the client, who has to handle a large number of smaller directly appointed specialist contractors
Suitable circumstances	1	Uncertain projects
	2	Large and complicated projects
	3	Major alteration projects
Most suitable projects	Major building projects such as city centre redevelopment and large commercial revenue earning buildings. The simple and direct legal relationships between specialist contractors and client favours "expert clients" such as major developers, who wish to save time, but need direct forms of redress should latent defects arise	

TENDER TYPES

In doing the deal for any transaction it is normal to want the best value for money. In this respect, placing building work is no different from buying food, kitchen equipment, cars or aircraft. However, for buying most goods, the object of the transaction exists in physical form, is produced in mass quantities and is identical, or near identical to other objects of the same type. This means that you can shop around for the best price, make a selection and offer to buy. Having a building built is different in that the object does not exist and usually will be produced specifically for the client to a unique design. It is rather like having a suit made to measure, but without the certainty that comes from a limited range of cloth and the limited range of sizes human beings come in.

It is usual, therefore, to arrange tenders for building work. Rather than the seller presenting its goods for purchase, the buyer advertises that it wants to buy. The principles of tendering are straightforward. Builders who wish to tender for work are presented with tender documents (sufficient information to help them decide). The documentation varies depending on the type of contract chosen and could range from a simple performance specification (as with pure design and build) to drawings and bills of quantities (as with traditional contracts with quantities).

Basic arithmetic is involved in deciding which and how many builders to invite to tender. The more you ask, the greater the competition and, therefore, the lower the price. This may be because there is a greater probability of inviting a very efficient builder, or a builder who is short of work and prepared to cut profit levels. It may also just be the result of a greater probability of finding a builder who has made a pricing error, which makes its tender lower. The last gives rise to the estimator's adage that "the biggest error gets the job". The disadvantage with accepting a tender based on erroneous pricing is that the builder may be tempted to cut corners or pursue extras in order to make up for the error. Having a large number of tenderers, conversely, gives more assurance that the price has been fairly obtained, without fraud. This gives comfort to those who are responsible for spending public money.

Tender strategy, therefore, involves a balance between competition and performance. Ranking tender options from the most competitive to the least, there are eight commonly used variants:

1. Open
2. Selective
3. Extension
4. Serial
5. Partnering
6. Single
7. Negotiated
8. Two stage

Open tendering

An open tender is open for all builders to enter, usually subject to paying a small sum for the tender documents. The invitation to tender is not normally completely open, as work will be categorised by type and size. It is advertised in a newspaper, journal or online, and on receiving the documents, the builder prices them in the form required. It then returns the tender itself, normally on a pre-printed form, or through a tender portal and before a time and date provided in the documents. As anyone could submit a tender, the client's advisors will carry out "post tender checks" to ensure that the tenderer is up to the quality of work required and sufficiently financially sound to last the duration of the project. Other factors may also be considered, especially the health and safety and quality assurance systems employed by the tenderer. Several tenders may, therefore, be excluded, where the builders do not pass the checks.

Open tendering provides the maximum competition for work, allowing new and inexperienced builders an opportunity to price. However, this may be at a greater risk of financial failure, or inadequate building standards. Open tendering is favoured by public authorities as it clearly indicates that justice has been done, without favour, in placing work. It is an expensive method of selecting a suitable builder as there is the potential for a very large number of applications to tender, but only one project to place. The tender ratio for the project (number of applicants to winning tender) is typically much higher than with alternative methods. Herein lays the contradiction with most forms of tender. As the competition is widened, the likelihood of a keener price increases but also does the cost of tendering! For reasons of better quality control, financial stability and lower tender costs, it is more usual to use a restricted tender than open tendering. However, open tendering remains as the reference base for placing work and is used regularly to feed into and support restricted tenders. For example, an open invitation will often be issued for builders to be added to a restricted standing list from which a selected number of firms will be chosen.

Open tendering – key decision factors

Advantages	1	Maximum competition, inducing lowest price "on the face of the tender"
	2	Gives new building firms an opportunity to tender – promotes new entrants
	3	Avoids restrictive and collusive practices – justice is seen to be done
Disadvantages	1	Greater risk of financial failure, or inadequate building standards from firms selected by open tendering
	2	The tender ratio for the project is typically much higher than with alternative methods, making the system expensive to the industry
	3	Carrying out post tender checks is administratively complex and time consuming

Suitable circumstances	1 As a reference base for placing work
	2 For high profile public works where no firms are seen to be favoured
	3 To feed into and support restricted tenders
	4 To allow new entrants to the market
Most suitable projects	Public sector works, possibly as the first part of a two, or more stage process

Selective tendering

In selective tendering, tenderers are drawn from a select group! For private traditional projects, the group often will be drawn up informally by the consultants employed on the project, particularly on recommendations from the architect and quantity surveyor. For public projects, a more transparent system is common, where builders are first invited by a public authority to apply for inclusion on a standing list. Admission to the standing list will be based on several criteria including financial standing, health, safety and quality systems used and evidence of adequate work in the past. Pre-tender checks will be carried out by the authority to ensure compliance with these criteria. A shortlist is then drawn from the standing list for the project being tendered. The shortlist should be drawn at random, but, as it is often difficult to get poor performers removed from a list, favoured builders emerge and a system of benign patronage similar to placing private projects may emerge.

Selective tendering is likely to increase the price level of individual projects for two reasons. Statistically, there will be fewer, but better organised builders in the competition and, thus less chance of an error winning the job. Also, the initial selection process will cut out less experienced, less financially stable or less capable builders, who nevertheless are cheaper. Selective tendering does, however, reduce the cost of tendering to the industry.

Selecting a limited number of builders to tender increases the risk of restrictive practices, such as manipulating the list to include a favoured builder, "price ringing", where the tenderers collude to produce a higher price, and "cover pricing", where the tenderers put in high prices rather than withdraw from the competition. The last tactic has the effect of giving the illusion of a higher level of competition than there really is. Two tactics used in avoiding some of these problems are to invite builders to tender who are outside the immediate local market (and thus less likely to be within a price ring) and to refer regularly to open competition.

Selective tendering – key decision factors

Advantages	1 Selection increases the likelihood that the chosen firm will be capable of carrying out the work in practical and financial terms 2 Selection reduces the tender ratio for contractors, reducing the cost of tendering to the industry
Disadvantages	1 Selection increases tender prices on any one project 2 Selection also increases the possibility of restrictive practices, further increasing prices and encouraging fraud 3 Maintaining standing lists of suitable contractors is administratively complex and time consuming 4 Carrying out "pre tender checks" also requires expert input 5 Getting on to tender lists may be difficult for new firms
Suitable circumstances	For most projects, particularly for clients who do not regularly build and who, therefore, do not have the opportunity to establish more continuous relationships with builders
Most suitable projects	Public and private sector projects of all sizes

Extension contracts

Both open and selective tendering suffer from the problem that the client does not know which builder it is going to end up with. Many ways of selecting builders work on the basis that you are better off with the "devil you know, than the devil you don't". Extension contracts seek to negotiate a further project on the basis of successfully completing an earlier one. The basis for negotiating the next contract is the prices provided in the last. These prices, for traditional contracts, might be from bills of quantities, with an addition for inflation and, perhaps, a reduction for more intensive use of site overheads.

Although extension contracts can be very useful to both client and contractor, they can be uneconomic to either depending on the state of the economy at the time. On a rising market, they will favour the client, as prices from the previous project will be keener, but on a falling market, they will favour the builder as prices from the previous project will be fuller. Therefore, negotiators for both sides need to have a view of the state of the wider market and economy whilst negotiating.

Extension contracts – key decision factors

Advantages	1 Allows longer term relationships to be developed between contracting parties 2 Maintains cost probity by using data from the initial contract(s)
Disadvantages	1 Can be expensive to either party depending on the state of the economy – on a falling market it is expensive to the client, on a rising market, to the builder 2 May produce higher prices due to reduced probability of favourable errors in tender
Suitable circumstances	The system can be considered wherever a client is currently using a satisfactory builder to carry out work and the existing contract can be used as a basis for further work
Most suitable projects	Additional projects on the same site or of the same type as recently constructed

Serial tendering

Many clients engage in programmes of work rather than single projects. Past programmes have included series of schools, hospitals and major roads. Serial tendering involves letting the first in a series on a competitive basis, but holding out the prospect of completing further projects in the series as extension contracts. Builders pricing for the first in the series will, therefore, price more keenly, especially if they think that fixed costs associated with plant and equipment can be spread over the whole series and therefore will work out lower per project. The follow-on projects are negotiated on the basis of the first in much the same way as an extension contract is negotiated from the original project prices. The advantages of serial contracts to the client are that they produce keener pricing, but also that they allow the overall performance of the builder to be evaluated in the first project before carrying on with the series. The advantage to the builder is that winning the first of the series holds out the prospect of a steady supply of projects. Further, performance and related profitability would be expected to continuously improve as expertise is gained during the series.

One major problem with serial tendering is that there is no legal obligation on a client to execute the whole or any of the series, other than the first project. The serial arrangement is merely an informal understanding that further work will follow. If the builder had made major investments on the prospect of lots of work, which does not materialise, the smaller series could be extremely unprofitable.

Serial tendering – key decision factors

Advantages	1 Allows longer term relationships to be developed between contracting parties 2 Encourages keen pricing as the tenderers will consider the whole series in pricing fixed-cost items. This should produce lower prices 3 Maintains cost probity by using data from the initial contract(s)
Disadvantages	1 The series of contracts are not contractually binding on the client and could be cut short in adverse circumstances. This could make the remaining contracts very uneconomical for the builder 2 Can be expensive to either party depending on the state of the economy – on a falling market the series may be expensive to the client, on a rising market, to the builder 3 May produce higher prices for individual projects due to reduced probability of favourable errors in tenders
Suitable circumstances	This is a suitable system wherever client organisations are considering a number of similar projects, either in terms of location or type
Most suitable projects	Serial tendering has been used for school, hospital and road building programmes

Partnering

As with extension contracts and serial tenders, partnering seeks to build on relationships between clients and builders, rather than use the impersonal commercial approach of open and selective tendering. A partnering arrangement will usually involve two parts, a normal building contract and the partnering agreement itself. In single project partnering, these two agreements will co-exist, with the partnering agreement asking the partners, for example to use their best endeavours to bring the project to a successful conclusion for all parties.

However, most partnering arrangements envisage "strategic partnering" or "framework agreements". These agreements are similar to serial tendering, where there are individual contracts for each project in a series, but there is also an overall partnering agreement. The partnering agreement essentially is an understanding that the series will be let to one builder on the basis of information provided in competition related to the first project or the series as a whole. Rather than simply basing the agreement of subsequent projects on prices in the first project, it is common to expect continuous improvement in key performance indicators over time. Thus, the fact that the builder is likely to get better at doing its job within the series is explicitly recognised. In addition to building relationships with a single builder, with its attendant possibilities of improving efficiency, framework agreements are efficient in that groups of projects can be let to a single builder without the need to engage in a protracted and expensive tendering process for each project. Once the builder is selected for the framework agreement, all subsequent projects are simply negotiated on the basis of defined criteria with an allowance for continuous improvement.

Partnering – key decision factors

Advantages	1	Introduces a goodwill element to building contracting, where parties will look beyond narrow commercial constraints to the benefit of participants as a whole
	2	Fosters a longer term aim of continuous relationships over a number of projects and attendant increases in performance, with increased profitability for all parties
	3	Framework agreements and single partnered projects can be let competitively, normally using selective tendering, and advantages are similar to that system
	4	Framework agreements reduce the cost and administrative complexity of tendering for individual projects as a whole series will be let on the basis of the award of the framework
	5	Encourages keen pricing as the tenderers will consider the whole set of projects in a framework in pricing. This should produce lower prices

Disadvantages	1 Separate partnering contracts, beyond the normal commercial building contract, are practically unenforceable and add nothing when relationships become strained 2 The set of contracts in a framework agreement is not contractually binding on the client and could be cut short in adverse circumstances. This could make the remaining contracts very uneconomical for the builder 3 It may be difficult to change a builder if projects within the framework are not going well, as opportunities should be given to improve performance 4 The method of placing individual projects by negotiation based on the framework may conflict with procurement legislation, requiring that clients obtain the most economically advantageous tender for each project[2] 5 Can be expensive to either party depending on the state of the economy – on a falling market, individual projects in a framework may be expensive to the client, on a rising market, to the builder 6 May produce higher prices for individual projects due to reduced probability of favourable errors in tenders
Suitable circumstances	1 This system is suitable wherever participants have the long-term aim of encouraging more continuous relationships. It is, therefore, relevant to expert clients with programmes of work 2 Framework agreements are suitable for clients with defined programmes of work, suited to wrapping in the framework
Most suitable projects	Any public or private sector project or series of projects, normally of larger size and where the client is engaged in building regularly or carrying out a substantial programme of work

Single tenders

In principle, there is nothing wrong with simply asking a single builder to price for carrying out a building project. The great advantage of a single tender is that you know the identity of the builder you will use. If there are very few firms in a locality capable of the type of work involved, a single tender may be inevitable, but the essence of the method is, first, that the invited firm does not know it is the only tenderer and, second, that, should it submit an inflated price the client will go elsewhere for other prices. The level of pricing and likelihood of obtaining better quotations elsewhere can be checked by a consultant architect or quantity surveyor to ensure that reasonable value for money is obtained.

Single tenders – key decision factors

Advantages	1 Advantageous where it is possible to get a price for building work, without disclosing that the tenderer is the only one under consideration 2 Quick form of tendering as negotiation is not envisaged
Disadvantages	1 Not competitive and may produce an expensive tender 2 Justice is not seen to be done with single tenders, so they are not suitable for public works
Suitable circumstances	1 Where very specialist work is in prospect 2 There is only one or very few suitable builders in the locality
Most suitable projects	Some types of very specialist work or work under very exceptional circumstances. Projects where the builder is known to have produced good work for that client in the past

Negotiated

A negotiated contract involves a single selected builder, but, in contrast with a single tender, selection is based on negotiation from the outset, rather than accepting a single price. There are two commonly used methods of negotiation, on the basis of out-turn prices, or on the basis of factor costs[3]. Whichever is chosen, thorough negotiation should produce an agreed price at a level reflecting normal pricing levels. However, the price will be higher than using tendering as there will be no "largest error winning the job", nor will the builder sacrifice normal profit levels to gain the work.

As with a single tender, negotiating means that the identity of the builder is known from the outset. However, even though this selection is effectively made, if only provisionally, at the outset, the need to negotiate in detail means that time is unlikely to be saved compared with obtaining competitive quotations. The time taken to negotiate and the need to keep parties at arms' length throughout the process provides few advantages over other methods of selection other than knowing who you are dealing with.

Negotiated contracts – key decision factors

Advantages	1 Knowing the identity of the builder under consideration 2 The prospect of negotiating a mutually acceptable solution
Disadvantages	1 Not competitive and may produce an expensive price 2 Justice is not seen to be done with negotiation, so it is not suitable for public works
Suitable circumstances	1 Where the identity of the builder is important, whatever the type of work 2 Where very specialist work is in prospect 3 There is only one or very few suitable builders in the locality
Most suitable projects	Some types of very specialist work or work under very exceptional circumstances. Projects where the builder is known to have produced good work for that client in the past

Two stage

Two-stage tendering is a hybrid combination of tendering and negotiation, in some respects similar to strategic partnering, but with only one project in prospect within the framework. For the first stage, it is usual to invite selected firms to tender on the basis of initial criteria. This could be on design, prices for key rates, construction management capability or a combination of factors. If more than one factor is being considered, a matrix might be drawn up with the factors scored against weightings. For the second stage, detailed negotiations will take place, but the basis for negotiations (e.g. using key rates, or an open book approach) will have been determined at stage one.

Two-stage tendering is very useful where it is desired to involve the builder early in the design-construction sequence. This might be for complex buildings where the design can strongly influence the construction methods used and vice versa. The builder can be appointed early on the basis of stage one and subsequently contribute buildability advice as the designs are developed. The builder can also start work well before the whole project is designed and costs can be agreed on a rolling programme basis.

Two stage tenders – key decision factors

Advantages	1 Allows the early appointment of the builder on defined criteria
	2 Allows the builder to contribute to design or early planning of work (buildability)
	3 Allows start on early sections of the project before detailed design of the whole
Disadvantages	1 It may be difficult to remove the builder if second stage negotiations break down
	2 The second stage is not competitive and may produce an expensive price
	3 Justice is not seen to be done with second stage negotiation, so it is less suitable for public works
Suitable circumstances	1 For complex single projects where construction advice may be needed before detailed design is completed
	2 Wish to save overall construction duration
Most suitable projects	Major complex landmark, commercial and industrial projects

SUMMARY

How building work is to be carried out and who is to do it are the two key strategic questions clients must ask at the earliest stages of a project. In asking these questions, they are presented with what appears at first glance to be a bewildering array of procurement options. Who to ask in order to get answers is also unclear – should one approach an architect, builder, quantity surveyor or project manager. Each might have a vested interest and push one option or another. However, it is possible to bring some order to the chaos by systematically analysing the strengths and weaknesses of each option and matching these to the nascent proposal.

It is possible to summarise the characteristics of procurement options in relation to the four key project objectives of quality, cost, time and risk mentioned at the start of this chapter. A simple decision table for contract types is illustrated in Table 1.1, with tentative solutions given in order of suitability.

A similar decision table can be constructed for tender choices, although factors of risk transfer or acceptance are not really relevant here. Table 1.2 summarises these choices:

Table 1.1 Decision table for contract options

Project objective		Procurement option (in order of effectiveness)
Maximise quality	1	Traditional with quantities
	2	Traditional without quantities
	3	Management contracting
	4	Construction management
Minimise time	1	Cost plus percentage, fixed fee or target
	2	Construction management
	3	Management contracting
	4	Design and build
	5	Design, manage and construct
Minimise cost	1	Design and build
	2	Traditional with quantities
Risk transfer to builder	1	PFI
	2	Design and build
	3	Traditional without quantities
	4	Design, manage and construct with guaranteed maximum price
Risk acceptance by client	1	Cost-reimbursement contracts generally
	2	Management contracting
	3	Construction management

Table 1.2 Decision table for tender options

Project objective		Procurement option
Maximise quality	1	Partnering and framework agreements
	2	Negotiated
	3	Single tenders
	4	Extension contracts
Minimise time	1	Two stage tenders
	2	Extension contracts (where relevant)
	3	Partnering (within a framework agreement)
	4	Serial tenders (within the series)
	5	Works orders (within measured-term contracts)
	6	Selective tenders
Minimise cost	1	Open tenders (for individual projects)
	2	Selective tenders
	3	Serial tenders

It is important to remember in considering a choice of procurement route that the suitability of any one route in relation to project objectives is tentative and more of a tendency than a prescriptive solution. There are large numbers of exceptions proving the general rule. Whatever route is chosen, detailed circumstances and the management abilities of key players often have an overriding influence. It is for this reason that apparently unsuitable procurement choices often work well and vice versa. A good choice, in theory, can sometimes be a disaster in practice.

NOTES

1. See Chapter 8.
2. For example, see The Public Contracts Regulations, 33 at: http://www.legislation.gov.uk/uksi/2015/102/introduction.
3. See Appendix to Chapter 2 for a detailed method of negotiating.

2

Building procurement procedures

AGREEING AND FORMING A BUILDING CONTRACT

What is a building contract?

Building contracts are, in principle, no different from any other contract for goods or services. There is no need for the contract to be in a particular form, or in writing. A verbal agreement formed by a householder with a passing tradesperson to repair a chimney stack is perfectly legal. The householder is inviting the person by simply having the stack there, the tradesperson makes an offer to do the work at a price and the householder accepts. The essentials of a binding contract are, therefore:

- **An invitation**
- **Offer**
- **Acceptance**
- **Consideration** (the payment from the householder and the work from the tradesperson)

However, not having the agreement written down could prove unsatisfactory for a number of reasons. The standard of work to be employed is not clearly set out, nor is the time that the person will take. Although you might not pay for the work until it is complete, you have no written record of the agreed price, or whether and what you would pay should more work be needed than is apparent at ground level. There would be considerable scope for argument, both as to the facts of the agreement and the underlying law. It is normal, therefore, to put building agreements in writing, both in relation to the detail of the work required and the terms of the agreement. As there are many similar building projects, the terms of the agreement are normally standardised into a pre-printed form. Using a standard form of building contract has many advantages including evidencing in detail the terms, allowing for a number of difficult circumstances (such as the need to vary the work involved), ensuring that the wording is tried and tested in use and in the courts and avoiding the cost and time associated with writing a purpose made form for every project. Modern standard forms of contract try to handle as many eventualities as possible within written terms, thus leaving as little as possible to interpretation (usually over a prolonged duration and at great expense) by lawyers.

However, standard forms of building contract have some disadvantages, not the least of which is the need to standardise procedures to go with the form. For example, using the JCT Standard Form of Building Contract SBC/Q2016 presupposes that tenders have been invited from builders on the basis of drawings and bills of quantities drawn up in accordance with the current method of measurement (New Rules of Measurement – NRM2[1]). If a different procurement path is chosen, then, both a different set of procedures and a different form of contract will be needed. In considering building contract formation, therefore, it is necessary to make assumptions as to the procurement path chosen and the form of contract being considered. It does not matter so much when dealing with broad principles, but there are considerable differences in detail within the large range of standard forms currently available worldwide.

One of the more widely used forms of contract in the United Kingdom, which also is the basis for many other forms internationally, is the abovementioned JCT Standard Form of Building Contract, with Quantities (SBC/Q2016). It is designed for traditional lump-sum procurement using consultant designers and reflects practice developed since the development of the modern building industry in the nineteenth century. As it is the most comprehensive of the JCT forms and reflects best practice, this chapter deals initially with contract formation for this form. Other widely used forms are the JCT Design and Build form (DB2016), intended for lump-sum projects where the contractor is to both design and build, and the JCT Minor Works Building Contract (with Contractor Design) MWD2016, intended for smaller lump-sum projects using consultant designers. Differences in practice with these forms are also considered here.

Forming the building contract – initial invitation to agreement

Traditional procurement using the with quantities form of contract, SBC/Q2016, envisages that consultant designers, under the direction of an architect and quantity surveyor, will produce detailed documentation in the form of drawings, specifications and bills of quantities, together with supporting information such as site surveys and soil reports. Some, but not all, of these documents are expressly referred to in the form of contract as contract documents and these form the substance of the contract. As they all contain cost significant information it is important that they are provided to the builders invited to tender for the project. The "Architect" and "Quantity Surveyor" are also named in the contract and, as the term "Architect" is reserved for registered architects[2], the alternative term "Contract Administrator" is provided for non-architect administrators.

Invitation – the tender documents

Tender documents where SBC/Q2016 traditional procurement is envisaged are tightly prescribed, effectively by the contract itself. The contract refers to the quality and quantity of the work being as described in the bills of quantities and that the bills have been prepared in accordance with the current method of measurement of building work (NRM2). NRM2 is, essentially an agreement between quantity surveyors and contractors' estimators, through their respective professional organisations, on what and how separate items of work in the bills of quantities will be measured. It is this, latter document, which details exactly what tendering contractors can expect when pricing the project:

1. **Tender drawings as delineated in NRM2.** Each work section of NRM2 lists the type and level of drawing to be provided with the bills of quantities. These will normally be quite detailed, but, as the builder will not have to actually build to the tender drawings, they may not be fully dimensioned working drawings nor may very large-scale detail component drawings be provided. Omitting some drawings is possible as the work will still be represented in written form in the measured items within the bills.
2. **Tender bills of quantities.** These will contain all the physical measured work items needed to build. These will be measured in accordance with NRM2. They will also contain a description of the general conditions under which the work is to be executed (the preliminaries) and specifications of the standard of work and materials required (the preambles). Incorporating preambles with the bills means that it is not necessary to provide a separate specification and SBC/Q2016 itself makes no reference to a specification document. As many firms of architects provide standard specification documents, where bills of quantities are to be used, these will either be physically incorporated into the bills or included by referring to them in the bills. Modern building projects also make extensive use of externally drafted documents such as national and international building standards and manufacturer's information. These will also be included in the bills of quantities by reference. NRM2 is included both in the bills and in the contract itself by reference.
3. **Supporting information.** Typically, the design team will carry out investigations to help them design the building. For example, in order to design foundations, it is common to carry out soil investigations. Reports of these investigations are helpful to tendering builders in designing temporary works (such as the base for a crane), or in pricing measured work (such as excavation). As determining the conditions under which work will be carried out is the builder's job, these reports are provided for information only, without liability, and will not form part of the contract.

4. **Invitation correspondence.** Covering letters, e-mails or web forms would typically contain instructions for submitting the tender (e.g. the time, date and method of submission) and required documentation (a standard form of tender and separate envelopes or web addresses for return of the tender form and priced bills of quantities).

Invitation – the tender procedure

Assuming that selective tendering is to be used for the project, suitable firms will be drawn by the client and design team from a standing or ad hoc list and a preliminary invitation will be issued to builders in order to gauge general interest. On expressing an interest, tender documents, supporting information and invitation letters will be sent to the tendering firms. Whilst the tendering builders will be busy pricing the bills of quantities, the client and design team have little function but to wait.

Offer – the offer documents

The offer documents returned by each tendering builder consist of the tender form with the written price for carrying out the project and the signatures of the authorised officers of the firm. If returned by mail, this would be in a sealed envelope containing the form. By e-mail, this could be by a simple reply or through a secure portal. It is normal to also call for the priced bills of quantities to be submitted at the same time, if by post also in a sealed envelope. If a builder's offer is under consideration, the corresponding bills will be examined in a process of due diligence to ensure that they have been calculated accurately and in accordance with the tender instructions. If an offer is not being considered, in order to maintain confidentiality, the envelope with the bills will be returned unopened, or the electronic submission will not be opened. There is no need to return tender drawings as these should not contain confidential material.

Offer – the offer procedure

The builders' estimators will price the bills of quantities in detail by calculating and placing a rate against each item (price per unit measure) and then calculating the product of the rate by the quantity to give the total cost of the item (the "extension"). Each section of the bill is then summed and sections are collected to produce the net cost of the offer. General overheads are either added to each rate or added as a lump sum. A further addition is made, after tender adjudication by the company directors, for the profit required on the project. Finally, authorised officers (usually two directors) of the company will sign and have witnessed the tender form, prior to submission to the client. Electronic submissions may involve secure keys held only by authorised officers, removing the need for physical signatures. In practice, as large amounts of building work are sub-contracted, sections of the bills of quantities are also sent out for sub-contractors to price, the builders' estimators only adding for overheads and profit before incorporation in the tender.

The tender

A tender (as opposed to tender documents) is a form of making an offer where all offers are submitted at or before a time stated by the client's advisors – for example by 12:00 noon on 26 October 2021. The tendering firms (also referred to as tenderers) will send by post, deliver by hand, e-mail, or submit through a secure portal, the tender form and bills of quantities. Once the time for submission has passed, the tender is closed. The client and advisors will open and consider the tenders.

Acceptance – the acceptance documents

It is a requirement for a valid contract to be formed that acceptance is unequivocal and does not qualify the accepted tender. Acceptance documents are, therefore, brief and usually consist of a simple statement of acceptance. However, JCT contracts envisage that a formal contract will be signed, indicating both acceptance and the basis of the agreement. There is often a gap between informal acceptance and formally signing the contract, but this rarely causes problems. Parties normally go on to formalise the agreement and it is only where an intervening event occurs in the gap that problems sometimes arise. Provided the informal acceptance was unqualified, should a problem arise, a contract will exist and the details of the agreement will be as set out in the offer documents – that is the tender form, completed bills of quantities, drawings and reference documents. This is likely to also be the case where acceptance is expressed as "subject to contract" in the event that the contract is never formalised.

Acceptance – the acceptance procedure

The client or its advisors (architect and quantity surveyor) will review all the tender forms submitted within time and arrange for them to be listed and formally recorded. This process will be carried out either in front of witnesses or by transparent electronic communication, to demonstrate fairness. The winning tenderer (for traditional contracts based on selective tendering, usually the firm submitting the lowest priced tender) will be informed and post-tender checks will be carried out. If these prove satisfactory, provisional acceptance will be given (often in the form of a "letter of acceptance") and arrangements will be made to formally draw up and sign the contract documents. Work starts after the contract is signed.

Complications

Before tendering

The process outlined above, where the tendering builders are pricing on the basis of tender documents provided by the client, places a great deal of emphasis on the completeness and accuracy of the documents. One way of expressing this is "if a requirement is not in the tender documents it is not in the project" and it is important that all variables that the client wants to include in the contract are presented to the tenderers in pricing. If something important is left out, the selected builder will justifiably be able to claim that it was not priced and does not form part of the agreement. Important variables presented in tender bills of quantities would include the start and finish dates of the contract, the interval for making any interim payments, the level of works insurance required and the amount of any liquidated damages. Most of these variables are listed in the articles of agreement and conditions of contract itself (e.g. in the Contract Particulars section of SBC/Q2016), so it is easier to remember to include them in the bills. In addition, the quantity surveyor will prepare bills of quantities in accordance with NRM2, which lays down information requirements of this sort in the Preliminaries Section 1.

Asking for alternative prices is not listed in the conditions of contract, so if alternative prices for project duration or a variable price as well as a fixed price are required, this should be stated in the bills. It is also common to ask for other special requirements, such as the provision of a bond to guarantee performance of the project (a performance bond[3]), or the execution of a collateral warranty[4] in favour of a tenant or eventual owner. Providing these extras can be onerous and expensive for a builder and will be refused or cause an extra cost if not made a requirement in the bills.

During tendering

ALTERATIONS BY THE CLIENT

It is often necessary to change details presented in the tender documents during the tender period. Items or materials may become unobtainable, or there may be a late change of mind by the client concerning some minor aspect. This can be allowed for, by communicating the change to all the tendering contractors. It is important that all tenderers confirm receipt of the communication. This ensures that parity of tendering is maintained between the tenderers during the tender period.

ALTERATIONS BY THE TENDERERS – NON-COMPLIANT BIDS

Changes are also often made by the tendering contractors themselves and, in order to maintain parity of tendering, these should not be allowed. It is usual to insert a clause in bills of quantities stating that changes will not be allowed and the original text will apply.

More drastically, a tenderer might be tempted to propose a different design for an element of a building, usually held out with the prospect of a lower price. For example, a tenderer might propose use of a steel rather than reinforced concrete frame at a lower price. Attractive though it might be for a client to accept the change of design (and the lower price), doing so would probably lead to objections from the other tenderers. They will consider that they should have also been offered the opportunity to price the alternative. To maintain parity of tendering, either the offer of the change should be rejected, or the change should be put to all the tendering contractors. It would still be open to the client to make the change as a variation once the contract was awarded. Knowing this, the tenderer concerned might price for the alternative, without putting it forward, in the hope that it will be accepted after work starts.

Having to maintain parity of tendering can, therefore, be seen as a constraint to innovative proposals from tenderers. This problem is caused by centralising design with the client and is reduced with types of contract based on design and build. For the latter, the tenderers can propose innovative designs at lower prices and, if they meet the brief, they can be considered in the tender.

After tendering

Once tenders are received, the major problems (assuming that selective tendering is used and all tenderers are considered suitable for the work) relate to errors and price manipulation. There are four broad issues to consider:

1. Errors on the face of the tender
2. Pricing errors
3. Cash flow pricing
4. Speculative pricing

ERRORS ON THE FACE OF THE TENDER

In pricing bills of quantities, estimators carry out a large number of arithmetical calculations involving multiplying rates by quantities and summing item totals, sections and summaries towards the tender sum. The potential for error, although much reduced by computer calculation, remains high. As most tenderers are pricing on very fine profit to turnover margins (often 1% or less), even a small error can have a large influence on the profitability of the contract.

PRICING ERRORS

A pricing error is not an error on the face of the tender. A tenderer may have put an apparently erroneous rate in bills of quantities deliberately. In contrast to errors on the face of the tender, pricing is internal to the tenderer and those inspecting the bills have no way of examining the intent of the estimator. However, a major pricing error is of interest to the client as it could endanger the financial stability of the tendering contractor. Correcting the error or allowing the contractor to withdraw from the tender might be in the interest of the client.

CASH FLOW PRICING

As most building work is paid for in instalments as work proceeds, there is a cash flow advantage in increasing prices for work carried out early in the contract. To maintain competitiveness, prices for work carried out later must be reduced, sometimes to the extent of not covering the cost of the work. The temptation to "front load" prices in this way, although good for the tenderer's cash flow, is not financially advantageous to the client. It also leaves the client exposed to extra expense should the contractor become insolvent during the later stages of a contract.

SPECULATIVE PRICING

Quantities of work often vary considerably during a contract from those originally envisaged at tender stage and differences in quantity are adjusted in payments to the builder. For example, excavation work may be more extensive than expected, or replacements in a refurbishment may be less extensive. The dynamic nature of the quantities gives opportunities for a tenderer to price speculatively. If quantities are expected to increase, a high rate or price will be inserted in the bills of quantities. To maintain competitiveness, lower rates will be inserted for quantities expected to decrease. Speculative pricing in this way requires considerable effort from the tenderer, but where margins are slight, may be worthwhile. For the client, the practice means that additional work will be more expensive than warranted and savings will also be less.

Tender reports

Post-tender checks will normally be carried out on the winning contractor itself, but with selective tendering these should be limited to ensuring the financial stability of the contractor. This would be done by asking for an up-to-date bank reference as well as an update of the company accounts. Checks on capability will have been already done as part of the pre-selection process.

However, a due diligence tender report will normally be prepared on the tender itself, including the detail of pricing in the bills of quantities, by the quantity surveyor. For a project where there is no quantity surveyor, the report would be prepared by the architect. Tender reports cover the issues identified above but would also contain other general matters of relevance to the client[5]. Typical contents include:

- An arithmetical check, where errors are described and the financial effect totalled up.
- A technical check for errors in pricing and price manipulation. The pricing errors are described and financial effect appraised.
- Comments on alterations to text and alternative details proposed.
- Comments on non-compliant bid aspects and elements.
- Appraisal of alternatives requested – for example alternative prices and project duration.
- Comparison with cost plan and earlier approximate estimates.
- Projected final account allowing for known or likely contingencies.

The arithmetical report would list the error, location and item, together with financial effect as an addition or subtraction, often in tabular format as illustrated in Table 2.1.

Table 2.1 Arithmetical report analysis

Page	Item	Add	Subtract
		£	£
	To summary		

Dealing with errors

The quantity surveyor or architect will have a duty of care in contract towards the client and should carefully analyse and report on the tender before recommending acceptance. He or she probably does not have a legal duty towards the winning tenderer. However, it is both morally correct and financially prudent to point out major errors on the face of the tender to the tenderer and expressly provide the opportunity for the tenderer to withdraw the tender. The same imperative does not apply to errors in pricing, as these are internal to the tenderer and may not be "errors" at all. A tenderer who does find such an error can still withdraw the offer as no contract has yet been formed. Good practice guidance in this area is found in the JCT practice note "Tendering"[6], where it recommends that tendering contractors are given a choice (in the tender documents) of *one* of two options.

Option 1 – Confirm or withdraw. The practice note indicates that should an arithmetical error be found in a tender, the client will invite the tenderer to confirm the tender as submitted or withdraw it. If the tender is withdrawn, the next most attractive offer would be considered.

Option 2 – Confirm or amend. As the next lowest offer might be higher that a corrected original tender, an alternative approach mentioned in the guidance is to present tenderers with a choice of either confirming the tender, or amending it (and taking the risk that the amendment will make the offer uncompetitive). The problem with this alternative is that competing builders very quickly learn (e.g. from common suppliers and sub-contractors) exactly who else is tendering and hence where they stand in a tender list. They can take a view on whether it is advantageous to confirm or amend the tender. The winner may amend the tender if it knows the amendment will not lose it the contract, thereby gaining a pecuniary advantage unavailable if it was forced to confirm or withdraw. For public contracts, where justice must not only be done, but be seen to be done, this option is not recommended.

Dealing with errors post-contract

Should a tender be confirmed, there will be an arithmetical mismatch between the rates and quantities in the bills of quantities and the tender figure. For example, an omission error in one section might have reduced the price of the tender by 5%, but all the other prices are correct. This means that, in relation to the overall tender figure, interim and final payments based on all other sections will be over-valued by about 5%. To correct this over-valuation, all interim and final payments will be adjusted downwards proportionately to the error.

Dealing with cash flow pricing and speculative pricing

It is unlikely that front loading bills of quantities would affect the order of a tender list unless prices were particularly close. In that case, a discounted cash flow calculation could be undertaken on the two lowest tenders to determine if the net present value of cash flow changed the order of preference. Similarly, a financial sensitivity analysis could gauge the effect of speculative pricing. The analysis would determine the extent to which quantities of vulnerable items will need to increase/decrease in order to change the order of a tender list. Overall, the object of considering the two issues is more to be forewarned of potential problems.

Reductions

Should the winning tender be in excess of expected costs (as reported, for example, in early approximate estimates), the client may be unable to proceed without reducing the scope of the work. A convenient way of reducing the scope is to prepare "bills of reductions" itemising work to be omitted or less expensive work to be substituted, priced at the rates provided by the winning tenderer. Typical items that might be reduced are landscaping, external work, finishings, expensive exterior cladding and fittings. The reduced scope and price will be put to the winning tenderer as a counter-offer. The tenderer is open to accept or reject the counter-offer and, knowing that it is in a strong negotiating position, might hold out for a more advantageous price. A common request is to ask for profit lost on the reduced scope of work, on the basis that the contract was priced for a particular duration at a particular profit level. If this loss of profit is conceded, the percentage profit on the contract will increase – in effect, the builder is being paid a profit on work *not* being done! For small reductions, this additional cost to the client might be acceptable, but should major reduction surgery be needed, it may be more financially advantageous to go out to tender again.

Reverse auctioning and preferred bidding

The JCT practice note "Tendering", in addition to the options on dealing with errors, contains advice on obtaining and placing tenders, largely following the procedures noted above. It also contains standard documents for inviting tenders and a pro-forma tender form. The use of internet-based tendering has brought to the fore reverse auctioning based on electronic submissions. The procedure itself predates the internet and involves approaching a tendering contractor and disclosing the best bid so far received. The target tenderer is then encouraged to undercut the bid. The lower bid is subsequently put to a second tenderer with the same object. Electronic communication has made it easy to formalise this process, with rapid bid-counter bid auctions. It is now expressly sanctioned by the Public Contracts Regulations 2015 (Part 2, Chapter 2, Section 4, Regulation 35)[7] provided that prior notice is given to tendering builders and the auction is based on clear criteria and a firm specification. It is envisaged that the auction would take place once formal conventional tenders had been received. A less acceptable form of auctioning is "preferred bidding" involving disclosing a winning bid to a preferred tenderer in the expectation that the bid will be matched. Although not illegal as such, preferred tendering could lead to a tenderer claiming through legal action the cost of preparing a bid, which was not seriously under consideration.

Although superficially attractive for a client, both practices have major drawbacks. An emphasis on grinding down prices by auction could, paradoxically, encourage cooperation between tendering contractors to maintain profit levels – a form of collusion that would ultimately raise prices. The assumption in an auction that price is paramount will encourage builders to skimp on quality and performance wherever there is ambiguity in any tender documentation. It could also encourage a "claims" mentality, where tendered prices are lowered in the expectation that margins will be increased post-contract by exploiting deficiencies in documentation, variations and delays. This goes against the spirit of cooperation engendered in the Latham[8] and other reports and puts great pressure on the design team to ensure that documentation is completely finalised before tendering. This cannot always be achieved or is deliberately sacrificed to save time.

Preferred bidding, particularly if a tender is only being used as a reference point for others, could have the effect of deterring tenderers from applying to certain clients. With smaller tender lists, competition will be reduced ultimately raising prices.

Formalising the contract

The contract documents

In JCT SBC/Q2016, the contract documents are defined in the form of contract as:

1. Agreement and contract conditions
2. *Contract* Bills of Quantities
3. *Contract* Drawings
   In addition, the bills of quantities will refer to other documents, including NRM2, national and international standards. This contract also envisages that portions of the work will be carried out to the contractor's own designs and there are further contract documents if this applies:
4. Employer's requirements for Contractor Designed Portion of the work (CDP)
5. Contractor's proposals for Contractor Designed Portion of the work
6. Analysis of the cost of Contractor Designed Portion of the work (CDP analysis)

Contractor designed work is considered more fully below in relation to procedures for design and build contracts. With the advent of building information modelling, a BIM protocol may also be incorporated as a contract document.

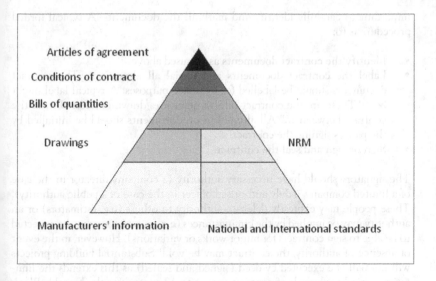

Figure 2.1 Pyramid of information forming the contract

There is no requirement as to the type and number of contract drawings. However, the bills of quantities contain the "quality and quantity of the works" (clause 4.1 of the contract) and contract drawings can be limited to layouts. If particular requirements must be met, in general it is in the bills of quantities that they will be found. Thus, requirements concerning performance bonds, insurance, attendance, warranties etc. will all be in the bills. These price sensitive items should already have been presented to the tendering contractors at tender stage – indeed, there should be no material difference between tender bills and contract bills other than the latter contain prices inserted by the tenderer.

One way of depicting the contract documents is as a pyramid of information forming the contract. The primary signatures are in the Articles of Agreement (bound in with the standard form), which refer to the Conditions of Contract. These govern the running of the contract and, in turn, refer to contract bills of quantities and drawings. Within the bills, reference will be made to a very wide range of (usually) standard documentation including NRM2, national and international standards.

The formalities

The object of formalising agreement to the contract, through contract documents, is to remove any doubt concerning what was agreed. It is, therefore, important to carefully identify and mark all the documents. A typical formal procedure is to:

- **Identify the contract documents** as itemised above.
- **Label the contract documents and initial all alterations.** The contract documents should be labelled for evidence purposes. A typical label might be – "These are the contract bills of quantities/drawings referred to in the contract between…." All alterations on documents should be initialled by the parties signing the contract.
- **Sign or sign and seal the contract**.

The signatory should have necessary authority (a company director in the case of a limited company, a duly authorised officer in the case of a public authority). These people may expressly delegate authority to others (e.g. estimators) or an authority may be implied by the circumstances (e.g. an estimator may be expected to be able to sign contracts for minor works or variations). However, in the event of absence of authority, the contract may be void! Substantial building projects will normally be executed by deed (signed and sealed) as this extends the limitation period for breach of contract from 6 to 12 years after the breach[9]. Blank forms for both simple and deed contract signatures are contained in all the JCT standard forms. In a paperless documentation regime, identification, labelling and signing can all be carried out electronically (e.g. by annotating PDF documents), but sealed documents will still need to be paper based.

The importance of formality

The formal identification of contract documents, initialling of alterations and signing (and sealing) of the contract has practical importance:

1. It is intended to avoid situations where there may be no clear acceptance of an offer, or acceptance has been made on the wrong terms (e.g. Contractor's standard terms rather than SBC/Q2016). JCT contracts envisage that the parties will lay out the documents and formally agree what is to be in the contract.
2. It clearly divides the project into two separate phases in dealing with changed work – pre- and post-contract. Changes that have been included on documents prior to signing the contract are part of the contract requirements. Changes that have not been included prior to signing can be dealt with as variations under the terms of the contract. (Discrepancies between documents and errors in quantities are also corrected and adjusted.)
3. It clearly evidences the contract and the contract documents. This reduces the scope for arguments about what was agreed.
4. Where a deed is required, it ensures that it is properly executed (and a 12-year limitation period is obtained).

MODERN PROCEDURES

A number of procurement innovations[10] have complicated tendering procedures relatively recently. The innovations include:

> **Private finance initiative**
> **Strategic partnering**
> **Supply chain collaboration**
> **Two stage, open book tendering**
> **Integrated project insurance**
> **Cost led procurement**
> **Greater use of design and build contract forms.**
> **Building Information Modelling**

Two common complications with these forms of procurement are an increased difficulty in comparing tenders in order to identify a "most economically advantageous tender" (MEAT) and an increased use of negotiation in agreeing the detail of projects.

Many of the above innovations ask for tenders on the basis of output measures (e.g. a price per bed-space for a hospital, or a price per student place for a school). Tenders may be in the form of a rent for a limited period after which the client will take over the building concerned. As overall life-cycle costs will depend on running expenses as well as construction costs, it is not possible to obtain the most economically advantageous tender on the basis of tendered first costs. Modern tender procedures may, therefore, include an assessment of:

> First cost
> Running costs
> Regular maintenance/Renewal costs
> Exceptional cost-in-use
> Cost/Benefit of design quality

In order to allow for these, a detailed assessment of each tender is necessary. To provide transparency in the tender process, invitations to tender will include a statement of the criteria for selecting contractors (based on the five points above). The process is known as "whole life cost assessment". The assessment is carried out as a post-tender procedure and is designed to avoid the need for post-tender negotiations, which may conflict with national and international procurement legislation.

Strategic partnering is a form of serial tendering where the initial contract or framework is set up in competition, with the selected tenderer being awarded further work based on their prices/method of pricing in the framework. This means that not every contract in a series will be tendered competitively and this has brought objections, again based on procurement legislation. Stringent regulations are in place for public works, which limit the size of frameworks that can be used without reference to competition.

Design and build contracts

Tender procedures for design and build largely follow those for traditional procurement, but the documentation and method of selection and placing contracts differs. Assuming that the JCT Design and Build Form, DB2016 is used, the contract documents consist of:

- Employer's requirements
- Contractor's proposals
- Contract sum analysis
- Articles of agreement and contract conditions
- If required and applicable a BIM protocol

Unlike the prescriptive nature of tender documents for traditional contracts, there is no indication of the nature and extent of the first three documents. The Employer's Requirements could be a brief written performance statement, sketch drawings, planning permission drawings, or detailed drawings and a bill of quantities. Contractors Proposals could be equally brief or extensive. The Contract Sum Analysis could be full bills of quantities produced in accordance with NRM2, schedules of rates, or a simple elemental cost breakdown. Should there be a conflict between employer's requirements and Contractor's Proposals, the latter prevail (Third Recital of the Articles of DB2016), placing the onus on the employer to make sure that the proposals fully meet the brief in the employer's requirements. The contract envisages that the employer will be advised by making express provision for an employer's agent. The agent (usually an architect, engineer, quantity surveyor or building surveyor) will advise on such matters as:

1. **Nature and extent of employer's requirements**. For example, performance criteria specified are accurately reflected in the requirements, that any drawn information is provided and that important constructional standards are specified. If warranties or novation of designers (see below) will be required, this should be included in the requirements. Either the agent or a separate Principal Designer[11] should produce and include pre-tender health and safety information in compliance with CDM regulations.
2. **Criteria for selecting contractors**. This will not usually be on price alone, but include design and specification details offered by the tenderers. Public clients have to show that they have obtained the most economically advantageous tender and may conduct a formal whole life cost assessment by

reducing qualitative criteria to present value costs. Whether or not this is done, the criteria for selection need to be determined and set out in the employer's requirements.

3. **Assessment of returned tenders**. This should be against the criteria published in the employer's requirements. The criteria should be scored, preferably in monetary terms, and clearly show the most economically advantageous tender. From this, a report will be produced, which includes comments on the adequacy of the contractor's proposals and contract sum analysis and on any discrepancies between employer's requirements and contractor's proposals. Once the requirements and proposals are adequate and properly match, the contract can be placed in the same way as for traditional procurement.

Novation of designers

Many regular clients prefer to retain their own designers but also wish to centralise responsibility with a single design and build contractor. They will thus include in the employer's requirements, the provision that the chosen contractor must employ the pre-contract designer as a condition of the award of the contract. The designer will have a provision in their contract with the client that they must agree to enter into a contract with the chosen contractor. The pre-contract designers will then work up their own design and deal with any post-contract design issues as the chosen contractor's sub-contractor. Whilst this is often an efficient and convenient procedure, it is unpopular with contractors. They are required to accept liability for the designer's work, but without control over its selection.

Minor works contracts using MWD2016

Tender procedures for minor works largely follow those for traditional procurement with bills of quantities, but as there are no bills of quantities prepared in accordance with NRM2, there is no defined set of tender documents. The contract envisages that certain documents may be provided by the Architect as contract documents including:

- Drawings
- Specification
- Work schedules
- Employer's requirements for any contractor designed portions

It also envisages that the Contractor may provide a Schedule of Rates. All of these are optional and reference to each may be deleted. This flexibility allows the documents to be fitted to the type of work – for example drawings may not be needed for renovation work and specification details could be included on the drawings for a very small contract. The documents are read together in determining the detail of the contract. For tender purposes, the architect would need to provide sufficient detail to allow builders to price the work, but where detail is not included, builders would be expected to fill in the gaps. In practice, many small builders provide detailed estimates itemising exactly what they will provide, where not delineated in the tender documents, in a similar way to contractors providing proposals with design and build contracts. As the detailed estimates will all differ in some respects, evaluation cannot always be on price alone.

APPENDIX – NEGOTIATING A TRADITIONAL CONTRACT

Once the decision to negotiate has been made, there are two methods of negotiating – on prices or on factor costs. For both methods, the same level of architectural detail is needed – that is sufficiently detailed drawings and specification to agree the price sensitive elements of the project.

Price-based negotiation

If negotiating on prices, bills of quantities will be needed to allow either a firm price to be calculated or the basis of pricing to be agreed. For the former, accurate quantities are taken out. For the latter bills of approximate quantities can be prepared and re-measured in the final account. The negotiation involves the contractor being asked to price the bills in detail, with the client's consultant quantity surveyor doing likewise (on the basis of price information from within his/her organisation or from published price sources). Rates for each item in the bill are then agreed. Overall, the measurement, pricing and negotiation are fairly laborious and time-consuming processes, often taking longer than obtaining prices in competition.

Factor costs-based (or "open book") negotiation

This approach recognises that most modern contractors use sub-contractors, bought in materials, plant and site equipment, rather than employing or engaging resources directly. If resources are bought in by a contractor, it is expected that some competitive process already exists for their selection. Consider the following breakdown of a typical contract sum:

Factor cost breakdown of a typical contract sum

Section	Percentage			Total
	Stated	Competitive	Negotiated	
PRELIMINARIES		5	7	12
MEASURED WORK				
Builder's own labour			10	10
Builders own plant and materials		20		20
Sub-contractors		40		40
Provisional sums	7			7
Overheads/Profit			11	11
Totals	7	65	28	100

The figure for "competitive" assumes a consultant (quantity surveyor or architect) insists on quotations to support pricing.

From the above table, it can be seen that only just over one quarter of a negotiated tender is truly negotiated. Factor cost-based negotiation can best be approached by agreeing the *basis* of pricing before commencing *detailed* pricing. To effectively do this, the contractor will be provided with sufficient documents to allow internal pricing, but this will stop short of initially providing bills of quantities. Negotiating documents provided for the contractor include:

1. A set of drawings including all available details in addition to those required by NRM2
2. Draft preliminaries based on NRM2
    2.1 Name of contract
    2.2 Names of parties
    2.3 Description of site
    2.4 Description of works
    2.5 Type of contract
    2.6 Details of alternatives and insertions in the Contract Particulars
    2.7 Anticipated and post contract programme
3. Anticipated total cost range
4. Anticipated pre-contract programme

5. Summary of contractor final selection process proposed
6. Criteria for selection

Information required from the contractor

1. Amounts for internal preliminary items (or a detailed statement of method of calculation)
2. Method of tendering bought-in preliminary items (such as site cabins, cranes and scaffolding)
3. All in labour rates required
4. Daywork rates required including overheads and profit
5. Sections of work to be placed with sub-contractors
6. Method of tendering sub-contract work
7. Method of obtaining prices for materials and plant
8. Amounts or percentages required for general overheads and profit
9. On agreement, priced bills of quantities matching the negotiated price based on NRM2 (for valuing interim payments and variations)

Armed with this information, there will be limited need for negotiation – agreeing some preliminary items, rates for labour for the few directly employed workers, additions for general overheads and profit. As the *basis of pricing* will be fixed for the other factors, work can be started before final agreement is achieved. In other words, the price can be progressively firmed up. Agreeing competitive factors (sub-contractor and materials prices, for example) will require an open book approach, where the consultant quantity surveyor can monitor appointments and purchases.

Appendix notes

1. Suitability of contractor for the work is assumed to be subject to a separate investigation.
2. All information received to be treated as confidential.
3. If a single contractor has been approached to negotiate, an adverse report can be made very early on and much time and effort can be saved.

NOTES

1. RICS (2012) New Rules of Measurement, NRM2: Detailed measurement for building works, RICS, London, UK.
2. Architects Act 1997, 1997 Chapter 22.
3. See Chapter 10.
4. See Chapter 8.

5. Further information on reporting is in RICS (2014) Tendering strategies, RICS, London, UK.
6. Joint Contracts Tribunal (2017) Practice Note – Tendering 2017, Sweet and Maxwell, London, UK.
7. SI 2015 No. 102.
8. Latham M (1994) Constructing the Team – Final Report, HMSO, London, UK, July, ISBN 011752994X.
9. Limitation Act 1980, 1980, Chapter 58.
10. Supply chain collaboration, two stage, open book tendering, integrated project insurance, and cost-led procurement were introduced in 2014 as a result of the UK Government Construction Strategy 2011. They are innovations linked to the promotion of Building Information Management and attendant cooperative methods of working. For BIM and the JCT contracts, see Chapter 15.
11. DB2016 assumes the Principal Designer will be the Contractor, but where substantial design is provided in the Employer's Requirements, the Principal Designer should, at least initially, be the pre-tender designer or Employer's Agent.

3

Interim payments

INTRODUCTION

It is traditional to make payments on account for building work, although for normal lump sum contracts this is not legally necessary. In the absence of express terms, payment would become due when the building was complete. However, three factors have led to the introduction of such terms:

1. **It is cheaper!** Making payments on account reduces the financing demands on contractors, thus allowing smaller low-cost firms without large amounts of capital to take on building work. This may reduce building prices in two ways:
   (a) Where the client can obtain finance at lower rates than the contractor, the overall cost of financing a project will be lower.
   (b) Removing the requirement to finance work means that a contractor does not need to have substantial fixed assets as security for loan finance. It can operate with low-cost rented accommodation, plant and equipment. As a consequence, however, building firms tend to be undercapitalised – they may have insufficient capital to deal with financial fluctuations.
2. **Contractors are unsecured creditors.** In the absence of payment on account, towards the end of a contract, a contractor would have large amounts of unsecured debt outstanding. This will be at risk in the event of the client becoming insolvent. Where contractors finance work, it is prudent for them to require financial security in the form of a charge on the property upon which they are building, or in the form of an insurance bond in their favour from the client to the amount of the contract sum. This would become payable on the liquidation of the client.
3. **It introduces flexibility into the contract!** It is often necessary for the client to change details of construction and this might increase the direct and indirect cost of the work. Although (in the absence of express terms) the contractor cannot demand payment on account for the work as ordered, it could for changed work. Allowing for such payment for all work gives the client the flexibility to make changes.

Note – Although the provisions of the UK Construction Act[1] have a requirement for interim payments, this is not directly a factor leading to their use. The
requirements merely reflect existing good practice and end-point finance is still
permissible under the terms of the Act (e.g. stages can be set at the interval of
the contract duration so there is just one stage due for payment at the end of the
contract).

End-point financing

Despite the norm of payments on account, it is not unusual for clients to require
the contractor to finance building work to completion. This provides:

- A natural incentive for the contractor to get on with the work.
- Conversely, the temptation to rush and skimp work!

However, end-point financing will normally only be offered by contractors where
they have control over design as well as construction. Having control over design
gives contractors the power to control many variations and avoid the problems
mentioned at (3) above. Variations could increase financing requirements and,
thus, affect the profitability of the contract. Procurement systems involving end-
point financing, therefore, are design/build based – pure design and build, design,
manage and construct (with or without a guaranteed maximum price).

Some forms of procurement such as PFI also require financing by the provider
(whether contractor or special project vehicle). Consequently, with these forms
of procurement, any client-driven design requirements need to be specified before
placing the contract as design control will be passed to the provider in order to
give it control over financing.

Stage payments

Some standard forms of contract (e.g. DB2016, Alternative A payment provisions)
make provision for payments on account based on reaching stages in the work.
Stages for a simple industrial building might be:

- Substructure complete
- Frame erected
- Cladding complete
- Interior complete
- External work and landscaping complete

Although addressing points (1) and (2) above (reducing contractor financing
requirements and lack of security for payment) and providing a natural incentive
for performance, stage payments are still problematic if changes are required by

the client (point 3). If changes are made between stages, immediate payment might be requested by the contractor and/or the next stage payment amended to allow for the change. If a large number of changes are made, this becomes administratively complex requiring the employment of a specialist to assess the amounts due at each stage and in the final account.

As a consequence, stage payments are really only suitable for relatively simple buildings, where few changes are likely. Simple commercial and industrial buildings (using DB2016) are examples of appropriate building types.

Measurement and valuation

The traditional method of assessing payment on account is by measurement and valuation. This method assesses the actual quantity of work executed by the contractor in a given time period (based on bills of quantities or a contractor's contract sum analysis), with an allowance for on- and off-site overheads and profit. By assessing work actually performed, it removes much of the direct financial incentive for performance but allows flexibility. Changes ordered by the client can be easily incorporated in the interim valuations as and when they arise.

This method was less prevalent, with the increase in contracting on design and build and PFI terms, but it is still widely used where traditional contract forms (SBC/Q2016, MWD2016) or prime cost assessment (PCC2016) are used.

Advance payment

The opposite of end-point finance is to pay for work in advance. This might be desirable for large pre-fabricated buildings or substantial elements of buildings where there is a large financing requirement for a contractor. There might also be a risk to a contractor should the building be cancelled or payment not be made. Advanced payment is also sometimes used to meet financing or tax deadlines – for example a capital expenditure tax break might expire before the building can be built and paid for, but payment in advance might be allowed by the Authorities.

Advance payment introduces the mirror problem for the client to that for the contractor with end-point finance. The client will require some security in the event of the contractor failing to perform. JCT SBC/Q2016 and DB2016 envisage the use of Advance Payment Bonds taken out by the contractor in favour of the client.

CONTRACT MECHANISMS FOR PAYMENT
SBC/Q2016

The rules governing payments on account in SBC/Q2016 are contained in Clause 4, with some reference to Clause 2.24/25. These reflect a very long-standing traditional practice of paying for building work by measurement and valuation on a monthly basis. The Architect, acting impartially between Employer and Contractor, is required to evidence the value of "**work properly executed**" in an interim certificate. Once issued, with a very limited number of exceptions, the Employer is required to pay the amount shown on the certificate. Amounts not due, perhaps because the work has yet to be carried out, or work is defective, are simply not "work properly executed" and do not form part of the certificate. The Employer does not, therefore, need to make adjustments to payment for most contentious issues. The regime for payment is fixed by reference to monthly **Interim Valuation Dates** (4.8 and the Contract Particulars). The expectation is that a valuation will be carried out (normally by the Quantity Surveyor – 4.9.2) at or around the Interim Valuation Date, the certificate will be issued within 12 days and the amount paid at the **Final Date for Payment**, within 9 days thereafter, giving a total maximum of 21 days from valuation to payment.

The period of 21 days from valuation to payment has changed little for many years but is now overlaid by largely unnecessary machinery invoked by the Construction Act[2]. The contract refers to the language of the Act – to "**Due Date**", "**Final Date for Payment**", "**Payment Notices**" and "**Pay Less Notices**", none of which are relevant if the above contractual duties of interim valuation and payment are carried out by the relevant parties. The Act is designed to stop payers from making unsubstantiated deductions from amounts due and is mainly relevant to contracts between contractors and sub-contractors and not main contracts. For most instances where the Act might be relevant, the Employer would already be in breach of contract for not following the traditional procedures.

There are very few amounts that an Employer could legitimately deduct from a certificate (the most common example being deductions for Liquidated Damages), and if the Employer does need to reduce the amount to be paid as stated in the certificate, then the Act requires that at least 5 days' notice prior to the Final Date for Payment be given in a **Pay Less Notice** and the notice must state the basis on which the deduction is made. For liquidated damages, the basis is plain as the amount of liquidated damages is stated in the Contract Particulars and the duration is clear from the dates of the Interim Valuation and the Non-Completion Certificate (2.31).

To reflect the common practice of contractor valuations, the contractor may submit to the Quantity Surveyor its own application for payment (4.10.1) and this would normally be incorporated in the Quantity Surveyor's valuation, be disputed, or modified. It is only if the Architect does not go on to issue a certificate of any description, that the contractor valuation trumps the Quantity Surveyor's and may subsequently form the basis for payment (4.10.2).

Two powerful sanctions in the contract cover late payment. First, late payment attracts simple interest at the rate of 5% over the official dealing rate of the Bank of England (4.11.6). Second, should non-payment of an amount beyond the Final

Date for Payment continue for seven days beyond the Contractor giving notice, then the Contractor can suspend work and claim the cost of so doing (4.13).

The Interim Certificate is based on the Gross Valuation of work done and is to include the following subject to a small retention (4.14.1):

1. Work properly executed
2. Site materials properly on site and adequately protected
3. Off-site materials provided that they are on a list submitted by the Contractor

The Certificate will also occasionally include additional items not subject to retention (4.14.2). A fairly detailed list includes such items as payments for tests, where the work involved is satisfactory (3.17), restoration costs under some options in the works insurance provisions (6.13.5) and loss and expense payments (4.20.1). The certificate will also occasionally deduct items not subject to retention, such as "Accepted defects" (2.38) (defects which have been accepted by the Employer in exchange for a reduction in cost), paying others to do work where instructions are not carried out (3.11) and the reduced value of non-conforming work (3.18.2).

The retention mentioned above may be deducted at 3% or a rate to be stated in the Contract Particulars (4.19), or the Contractor may provide an insurance backed retention bond instead (4.18). Where retention is held, it is to be in trust and put into a separate Employer bank account if so requested by the Contractor (4.17.3) (this is to protect the amount in the event of the Employer becoming insolvent). The retention percentage is halved after Practical Completion has been reached (4.19.2) and no retention is held once defects have been put right at the end of the project (implied in 4.19)

Materials on and off-site

There are special provisions dealing with materials yet to be incorporated in the works. Materials would normally remain the property of the Contractor until fixed in the works. Clause 2.24 makes it clear that they may only be removed from site with the Architect's consent and once paid for become the property of the Employer[3].

Prefabricated **materials off-site** are required to be on a list in accordance with (4.16) and once these are paid for, they also become the Employer's (2.25). Clause 4.16 provides that listed off-site materials may be of two types, "not uniquely identified listed items" and "uniquely identified listed items". Both types must be:

- Owned by the Contractor before payment by the Employer (4.16.2.1)
- Adequately insured (4.16.2.2)
- Set apart and clearly marked for the Contract/Employer (4.16.3)

- For "not uniquely identified listed items" backed by a suitable bond in favour of the Employer (4.16.5)
- For "uniquely identified items" backed by a suitable bond in favour of the Employer if this is stated as required in the Contract Particulars (4.16.4)[4]

DB2016

Procedures with this form of contract are much the same as with SBC/Q2016 except that there is no certifier and the contractor makes an **Interim Payment Application** (4.7.3) based on work carried out up to the **Interim Valuation Date** not more than 7 days earlier. The payment will be (with alternative A – stage payment) at the completion of a stage of work (4.12.1) or (for alternative B – periodic payments) on a monthly basis (4.13.1). Payment becomes due 14 days after the Interim Payment Application (4.9.1) giving, as with SBC/Q2016, 21 days from valuation to payment. Late payment attracts simple interest as with SBC/Q2016 (4.9.6).

There is no named quantity surveyor, or architect to independently moderate the application and, if the Employer disputes an amount (perhaps, for example, for defective work), it will need to operate the Construction Act machinery to issue a **pay-less notice** and reduce the amount of the payment. The Interim application is based on the Gross Valuation of work done. In alternative A this is based on the cumulative value of stages (4.12) and in alternative B on the value of work carried out so far (4.13).

MWD2016

The general approach of this form is similar to SBC/Q2016, but simpler. The Architect issues Interim Certificates on a monthly basis, including the value of work up to the **Interim Valuation Date**, 7 days earlier. The amount is based on what he/she considers to be the total value of work properly executed and materials on site. Payment becomes due 14 days after the issue of the certificate (4.3) again giving a 21 day valuation to payment duration. Late payment attracts simple interest (4. 6).

There is no separate named quantity surveyor. The expectation is that the Architect will fairly value the work and the Employer will not dispute this. However, if the Contractor is dissatisfied with the amount of a certificate, it can only commence adjudication proceedings.

Valuation practice

Interim Certificates and valuations for JCT forms are always calculated gross with a deduction at the end for work previously executed. For SBC/Q2016, valuation and certification are clearly separated and the named Quantity Surveyor is responsible for carrying out or approving the valuation. Where there is no QS, the following procedure would be best practice, either for the architect where there is one or the contractor. The valuation will normally follow the sequence of the Bills of Quantities fairly closely and contain the following headings:

1. **Preliminaries**. These are usually assessed on one of three bases;
    (a) Time based. The total value of preliminaries is divided by the number of weeks in the contract and then multiplied by the number of weeks since the contract started.
    (b) Value based. The total value of preliminaries is calculated as a percentage of the measured work in the bills of quantities and this percentage is applied to the value of measured work executed to date.
    (c) Based on an estimate of how preliminaries are actually accrued during the contract. The Quantity Surveyor and Contractor's commercial manager together agree a time/cost schedule and assess preliminaries accordingly.
    The last method (c) is most accurate and preferred for larger complex contracts. It helps the contractor, by reflecting the high cost of setting up at the start. For smaller works, (b) is simple to operate but will considerably understate the cost of early work and (a) might be preferred for this reason. If (a) is used, it can result in overpayment if the contract is running behind time. It should, therefore, be coupled with a review of progress to ensure that delay is not leading to overpayment.
2. **Measured work**. This is valued by assessing the quantity of work done against the bills of quantities items and rates. In traditional "trade-order bills", this is fairly straightforward as the trades follow the sequence of work.
3. **Variations** are measured, valued and included as measured work.
4. **Site materials** are assessed by the Quantity Surveyor, usually assisted by the Contractor, who may provide a list of materials and supporting invoices/delivery notes. The materials are required to be (4.14.1.2):
    (a) Properly protected
    (b) Not on site prematurely
    Note – The materials do not need to be owned by the Contractor when paid for (2.24, 4.14) so supply conditions from materials suppliers should be checked to ensure that they do not contain "reservation of title" clauses.
5. **Off-site materials**. These are included if they are on the lists mentioned above and comply with the conditions (i.e. are properly set aside, insured and bonded).
6. **Statutory fees and charges**. Amounts payable to "Statutory Undertakers" etc. are reimbursed to the Contractor without deduction for retention.
7. **Agreed claims**. Amounts for loss and expense that have been accepted by the Architect are payable in Interim Certificates without a deduction for retention[5].

8. **Fluctuation payments**. If allowed by the contract, these are added as they accrue.
   (a) **Formula-based payments** are subject to retention.
   (b) **Traditionally assessed payments** are not[6].
9. **Retention**. This is deducted on items (1), (2), (3), (4), (5), and (8a) above at the "retention percentage". Once the works are complete (Practical Completion), this percentage is halved. Retention is not held if the Contractor provides a retention bond executed in favour of the Employer.

Separate issues

ADVANCE PAYMENT (4.7)

Where an advance payment has been made to the Contractor, this is deducted in accordance with the terms of the Contract Particulars (from an Interim Certificate – 4.15.2). The Quantity Surveyor will advise that this deduction is due, but, as the adjustment is made by the Employer, not make an allowance in the valuation.

DEFECTS

The Interim Certificate is expressed to include "work properly executed" (4.14.1.1), but the Quantity Surveyor is not qualified or responsible for approving the quality of work. The Architect will be given the opportunity to deduct from the valuation for defects prior to certifying the amount due. This opportunity will be expressly provided at the foot of the Quantity Surveyor's Valuation Statement.

LIQUIDATED DAMAGES

These are deducted by the Employer, so are not covered in the Interim Certificate or Valuation. The Architect will advise the Employer separately of the right to make this deduction from the payment, subject to Construction Act provisions, involving issuing a "Pay Less Notice" (4.12)[7].

INSURANCE

If the Contractor fails to insure the works, the Employer may insure and deduct the amount from sums due. Again, this is not covered in the Interim Certificate/Valuation. The Architect will also advise the Employer separately of the right to make this deduction (6.12.2).

Valuation practice for DB2016

The Contractor applies for payment with this form and where provision is made for stage payments, the value of the stage, as noted in the contract sum analysis, is merely added when it is reached (subject to any adjustment caused by variations, claims etc.). Where valuations are to be on the basis of periodic payments, some form of price analysis of the contract sum needs to be available as the basis for measurement and valuation. This is expected to be a contract sum analysis, but unlike bills of quantities with SBC/Q2016, the form of this instrument is not defined. The analysis will be most useful when it is full bills of quantities and some Employer's Requirements include a requirement for production of bills post-tender, either by the contractor or a consultant. Alternatively, the Analysis should at least contain schedules of rates following the items in NRM2. The valuation can then follow the procedure above.

The contractor should only apply, in the interim applications, for work properly executed (4.13.1.1) and materials properly brought to site (4.13.1.2). However, as there is no separate certificate issued by an architect or similar professional, any reductions for defective work, or materials brought to site too soon, would have to be made from the Employer's payment. The reductions will be governed by the Construction Act notice requirements, which accordingly assume greater significance with this form.

Valuation practice for MWD2016

The Architect is responsible for valuing work executed and materials on site, but, the valuation instrument is not defined. Schedules of Rates and Work Schedules are envisaged by the form and these may be used for interim valuations. If schedules of rates are provided, work can be valued on the basis of these in much the same way as with bills of quantities. Work schedules loosely itemise work (generally of a repair and maintenance nature) and may be priced per activity by the builder. They can thus also act as a suitable stage-based valuation instrument. For new work, it is common for Architects to request (or builders to suggest) a cost breakdown on the basis of stages of work. Payment is then made on a stage basis in the same way as envisaged in Alternative A of DB2016.

For all methods of payment, it is common to request that the builder makes an interim application for payment, with the architect approving the figure and issuing a certificate. The architect will make reductions on the certificate for work not properly executed and materials brought on to site too soon. These reductions will not, therefore, be subject to the requirements of the Construction Act.

See attached sample valuation and preliminaries breakdown.

ANNEX A – A BREAKDOWN OF PRELIMINARIES

| Item | Set-up | Running | | | Remove | Total |
		Per week	Weeks	Total		
Progress chart	100					100.00
Person in charge		500	30	15,000.00		15,000.00
Plant etc.	2,000.00	500	12	6,000.00	1,000	9,000.00
Safety, health etc.	2,000.00					2,000.00
Temporary protection	1,200					1,200.00
Temporary storage	500	100	8	800.00	200	1,500.00
Temporary screens	600				400	1,000.00
Scaffolding	3,000.00	500	18	9,000.00	1,000	13,000.00
Cleaning/ Disposal					2,000.00	2,000.00
Telephone		100	16	1,600.00		1,600.00
Insurance	5,400.00					5,400.00
Electricity		40	20	800.00		800.00
Total	14,800.00			33,200.00	4,600.00	52,600.00
					Check total	52,600.00

ANNEX B – A SAMPLE VALUATION

Job	Twelve flats at Deepdene			
Job no.	1234			
Date on site	20/06/2020			
Valuation	6			
			£	£
Bill 1 preliminaries	As schedule			
	Set up		73,350.00	
	Running			
	Weeks	25.00		
	Per week	4,050.00	101,250.00	174,600.00
Bill 2 substucture	Complete		327,600.00	
	Less			
	BQ page 32/G	30,000.00		
	BQ page 32/H	12,000.00	42,000.00	285,600.00
Bill 3 concrete work	BQ page 35	90,970.00		
	BQ page 36	75,447.60		
	BQ page 37	73,806.00		
	BQ page 38	23,138.36		
	BQ page 39	6,618.00		
	BQ page 40	25,926.00		

	BQ page 41/A-K	1,980.00		
	BQ page 42/F	432.00		
	BQ page 42/G	72.00		
	BQ page 45/M	1,638.00	**300,027.96**	**300,027.96**
Bill 4 Masonry	BQ page 47/A-D		**1,200.00**	**1,200.00**
(etc.)				
Variations additions				
AI 3 Swimming pool	Final A/C item 6		98,338.00	
	Less to complete		18,000.00	**80,338.00**
AI 5 Drainage remeasure	Final A/C item 8			
	Excavation	7,422.00		
	Surrounds/Beds	17,994.00		
	Pipework	5,058.00		
	Chambers say	13,200.00	43,674.00	**43,674.00**
Materials on site	Invoices for			
	Bricks	8,538.00		
	Blocks	1,650.00		
	Sand	618.00		

	Cement	1,362.00		
	Aggregates	2,940.00		
	Chamber covers	3,000.00		
	Timber	4,200.00		
	Pipes	1,200.00		
	Roof tiles	6,000.00	29,508.00	**29,508.00**
Fluctuations	**See separate list**			**1,454.00**
Gross valuation				**916,401.96**
Retention		3.00%		**−27,492.06**
Net valuation				**888,909.90**
Previous				**−726,216.00**
Valuation 6				**162,693.90**

NOTES

1. Housing Grants, Construction and Regeneration Act 1996 – Part 2 as amended by the Local Democracy, Economic Development and Construction Act 2009 – for details of the Construction Act, see Chapter 14.
2. See footnote 1.
3. This provision may conflict with terms in sub-contract and supply contracts leading to problems of ownership of materials in the event of the Contractor becoming insolvent – see Chapter 7.
4. For more on bonds, see Chapter 10.
5. Further detail on claims is given in Chapter 6.
6. See Chapter 11 for methods of assessing fluctuations.
7. See Chapter 14 for details of the Construction Act.

4

Final accounts

INTRODUCTION

Express and implied variations provisions

The final account is the amount payable to the contractor after the completion of the works in accordance with whatever contract is being used. It reconciles the amount finally paid with the initial contract sum. The main reason for a difference between the two is variations in the design of works after the contract is agreed. Although SBC/Q2016 allows for variations and states (3.14.5) that no variation "shall vitiate this contract", there is a question of whether all variations are permissible. This gives rise to a distinction between "variations to the contract" and "variations to the works".

Variations to the contract involve varying the terms of the agreement itself. In a simple contract (without express variations provisions), varying the works is a variation to the contract and will, in effect, require a separate agreement (a mini supplementary contract) for each variation. Variations to the works are allowed within the terms of most standard forms of contract. For example, SBC/Q2016 allows for variations in Clause 5.1, MWD2016 in Clause 3.6, and DB2016 in Clause 5.1 (variations are called changes in this contract, but "a rose by any other name would smell as sweet!"[1]). These variations are not, therefore, variations to the contract.

Problems occasionally arise when very large or numerous variations are ordered, which may not, when viewed objectively, be variations to the works but change the agreement itself. Take the example of an architect purporting to change, by variation, a contract for building a concrete framed office block to that for two houses. A court may well decide that this change is beyond an **implied term** related to the extent of a variation and refuse to support the action of the Architect even though the wording of the contract allows the variation.

The question of whether there is an implied term limiting variations and what the limit is has been put to the courts on several occasions, particularly when working to contract rates, is extremely uneconomic for a contractor. A decision that an implied term has been breached might allow a contractor to recover costs for changes on a cost recovery basis rather than at contract rates. It is important, when considering variations clauses and many other clauses in printed forms of contract, to remember that the wording is often circumscribed by implied terms limiting the applicability of the literal meaning. In dealing below with variations and final accounts, only the express terms of the contract are being considered and for nearly all variations this will be adequate.

To some extent, for public contracts, the limits of variations have been codified by the Public Contract Regulations 2015[2], where regulation 73.1(a) requires termination where "the contract has been subject to a substantial modification which would have required a new procurement procedure" – this is allowed for in the termination provisions of all three contracts (8.11.3 for SBC/Q2016, DB2016, 6.10.3 for MWD2016).

JCT SBC/Q2016 contract provisions

Assuming JCT SBC/Q2016 is the contract in use, the following are the main clauses bearing on the final account.

Architect's instructions (3.10–3.14)

These clauses are enabling provisions, allowing the Architect to issue instructions and requiring compliance. Instructions are the vehicle by which the works are varied or other extras/savings are authorised. In essence:

1. The Architect can issue instructions (an **Architect's Instruction**!) (3.10–3.14).
2. The Architect can require compliance within 7 days or (through the Employer) get others to do the work and deduct the cost in the final account (3.11).
3. Instructions are to be in writing, but not in any particular form (i.e. drawings, agreed site minutes, letters and standard instruction forms can all be instructions) (1.7.1).
4. Verbal instructions are of no effect but must be either:
   (a) Confirmed by the Contractor to the Architect within 7 days. If the Architect does not dissent within a further 7 days, the confirmation becomes an instruction (3.12). (In practice, this is a powerful tool for the contractor, in the common situation of trying to get confirmation for large numbers of verbal site instructions.)
   (b) Confirmed by the Architect up to the issue of the final certificate (3.12).
5. The contractor can question the validity of an instruction (i.e. whether the instruction is empowered by the contract). This is done by requiring the Architect to specify the provision empowering an instruction and allowing immediate adjudication if there is still a dispute (3.13). Matters normally not

empowered by the contract relate to detailed methods of working adopted by the Contractor.

6. The Architect may issue instructions requiring a **variation** (3.14) (in the case of a Contractor's Designed Portion, limited to a variation of the Employer's Requirements).

Note – there is no role in the SBC/Q2016 instruction procedures for instructions from such specialists as structural engineers, clerks of works, or quantity surveyors. If they issue instructions:-

1. They can be ignored.
2. They must be confirmed by the Architect.

This restriction is designed to centralise responsibility for instructing the contractor on one person. However, to allow for the smooth running of a project, the contractor will in good faith often act on verbal instructions from such people. Clerks of Works have a special mention in (3.4), where their instructions are specifically mentioned as having to be confirmed in writing by the Architect within 2 business days.

The Employer is entitled to know that payments have only been made in accordance with the contract. Thus, consultants (particularly the Quantity Surveyor) and the Contractor have to show that the form for the issue of instructions outlined above has been complied with. Public clients will normally expect explicit evidence against final account items of the source authority for the item (e.g. a written instruction, a drawing reference or an agreed site minute), but private clients can demand the same evidence. Thus, proper cross-referencing of authority is important.

Contract Bills and contractor's design portion documents – errors etc. (2.14)

Quantities are guaranteed in the Contract Bills and errors are treated as variations to the work, not requiring an Architect's Instruction. However, as they can cause considerable embarrassment to the Quantity Surveyor and he/she prepares the final account, they rarely are seen as final account items! Errors here relate to errors in quantity and not to errors in pricing, which will remain uncorrected.

JCT SBC/Q2016 makes provision for the Contractor to design portions of the works (e.g. the frame, a lift installation or heating system). The contract treats these portions as "mini design/build" contracts within the main form and the wording and process for them is very similar to the full DB2016 form considered below. Only errors in the Employer's Requirements for this Contractor's Design Portion (CDP) work would entitle the contractor to payment. Errors in the Contractor's Proposals may be corrected, but without financial addition.

Variations (section 5)

Section 5.1 defines "Variation".

Section 5.2 gives three methods for valuing variations.

1. By direct agreement between the Employer and Contractor (5.2.1).
2. The traditional process of measurement and valuation by the consultant Quantity Surveyor in accordance with "valuation rules" (5.2.1). The process encompasses valuing variations to a CDP in line with a CDP analysis (5.8).
3. Provision of a prior quotation by the contractor, which is agreed with the consultant QS (A "Schedule 2 Variation Quotation"). This bypasses detailed measurement by allowing the contractor to quote for work. The Schedule 2 Quotation still has measurement and valuation to fall back on, should the quote prove unacceptable. The procedure is most useful for discrete items distinct from the main work – for example for the addition of a tennis court to a project for constructing an apartment building.

DIRECT AGREEMENT

This provision allows the contracting parties themselves to take a business approach and short-circuit the detailed process of measurement and valuation or quotation and agreement.

THE TRADITIONAL PROCESS

Traditionally, variations in JCT contracts attempt to tie the cost back to the original agreement as much as possible. In essence, on entering the building contract the parties are agreeing both to work to a contract sum and to work on the basis of the contract sum in costing variations. Accordingly, where additional work is required, adjustment of the cost is made on the basis of the rates inserted in the bills of quantities when they were initially priced. This would be so even where the initial rates are erroneous. The contract puts into effect this approach in Clause 5.6.

5.6. Defines the Valuation Rules for valuing the cost of variations:

1. Contract Bill rates are used for omissions and additions wherever possible (to show compliance with this clause, the Quantity Surveyor will cross-reference final account items to the bills) (5.6.1.1, 5.6.2).
2. "Pro-rata rates" are used if exactly equivalent items cannot be found in the Contract Bills. Adjustments are made as appropriate to suit changed conditions etc. (5.6.1.2).
3. A fair valuation is made for work not appearing in the Contract Bills (5.6.1.3). A fair valuation might use rates for similar work for the same

contractor on a different project or might be based on published building price books or on-line data.

4. Approximate quantities are to be valued on the basis of approximate quantities rates where the quantity is a reasonable forecast of those actually used (pro-rata if not) (5.6.1.4–5.6.1.4.5).

5. If it is not possible to use measurement and valuation, then Dayworks rates may be used. Dayworks might be used for remedial work and alterations, but not normally for new work (5.7).

6. Re-rating may be allowed where work not varied is so affected by a Variation as to increase costs. The opposite (i.e. extra work making remaining work cheaper) would not normally allow re-rating downwards. This is because the source of the variation is the Employer's agent, the Architect – effectively unilateral action (5.9).

Variations are valued by the Quantity Surveyor (5.2.1) and the Contractor is given the opportunity to be present (5.4).

SCHEDULE 2 VARIATION QUOTATION

The Architect may request that the contractor provides a quotation for carrying out a variation (5.3.1) and, unless the contractor objects within 7 days, the procedures in 5.3 for a Schedule 2 Variation Quotation will then apply. The contractor is required to submit a quotation containing:

1. The quotation (Schedule 2, Clause 1.2.1)
2. Extension of time required (Schedule 2, Clause 1.2.2)
3. Loss and expense required (Schedule 2, Clause 1.2.3)
4. Cost of preparing quotation (Schedule 2, Clause 1.2.4)
5. Resources required (Schedule 2, Clause 1.2.5.1)
6. Method (Schedule 2, Clause 1.2.5.2)

The Architect, on behalf of the Employer, can accept the quotation in accordance with Schedule 2, Clause 4, but this is expressed in the terms of counter-offer proposals of an adjustment to the Contract Sum and revised Completion Date. For the quotation to be binding, the Contractor would need to confirm acceptance of these proposals. If the quotation is not accepted by the Architect, then the rules for traditional valuation are to apply. The Contractor may claim the cost of providing a quotation if it is rejected (Schedule 2, Clause 5.2)[3].The object of asking for an estimate for any extension of time (2) and loss and expense (3) is to pre-agree the knock on effect of the variation on the programme and indirect costs such as site overheads. This will avoid the possibility of a claim arising related to these matters[4].

Contract sum and adjustments (4.1–4.4)

These clauses summarise the procedures for adjusting the Contract Sum and the items to be adjusted. They give in a conveniently collected form a guide to all the clauses throughout the contract affecting the final account, although the term "final account" is not mentioned.

Overall lump sum adjustments to the Contract Bills are adjusted proportionately to the value of measured work on each variation item (5.6.3.2). Preliminary items affected are also adjusted (5.6.3.3), except for preliminaries on items set against "Provisional Sums for defined work" (see below).

The final certificate (4.26)

The Architect is required to issue a final certificate not later than 2 months after the latest of:

1. The end of the "Rectification Period"
2. The issue of the "Certificate of Making Good Defects" under 2.39
3. The delivery of the final account under 4.25

The certificate is a statement of the amount finally owing to (or owed by) the Contractor, subject only to any directly set-off amounts by the Employer. Setting off amounts (which would relate to such sums as liquidated damages) is only allowed subject to clause 4.12, which implements the Construction Act provisions for set off. The date of issue of the final certificate is the "Due Date" for final payment in accordance with the Act and the "Final Date for Payment" is 14 days thereafter. The Employer must give a "Pay Less Notice" in accordance with the Act at least 5 days before the final date for payment.

THE EFFECT OF THE FINAL CERTIFICATE (1.9)

The issue of the final certificate has the effect of:

1. Evidencing that the Architect is satisfied with the quality of materials and standard of workmanship (where such satisfaction is required in any Contract Document).
2. Evidencing that the contract has been complied with in calculating the final amount due to the Contractor.
3. Evidencing that all extensions of time to complete the works have been given.
4. Evidencing that all loss and expense claims have been settled.

The implications of these points are that if the Employer has a dispute with the Contractor concerning matters that require the Architect's satisfaction, or judgement in relation to quality/extensions of time, then it has no recourse under the Contract. The same applies if the Quantity Surveyor has overpaid the Contractor. This does not apply if the Employer instigates the dispute resolution mechanisms within 28 days of the issue of the Certificate.

Normally, the Employer would have 6 years in a simple contract and 12 year for a deed contract to take action against the Contractor. Other routes to redress in such disputes could be by taking action in Tort or taking action against the Architect or Quantity Surveyor for negligence in performing their duties.

The practical objection to 1.9 from the point of view of architects is that it is almost impossible to check every detail of the work for "Satisfaction". They may be open to negligence claims, when the culpable party is really the contractor. To avoid liability for wrongly expressing satisfaction of work, requirements should, therefore, be expressed objectively in terms, for example, of national and international building standards, rather than to the subjective opinion of the Architect.

JCT DB2016 contract provisions

The JCT Design and Build contract (DB2016) also make provisions for settling the final account, but the approach and terminology are different. In design and build contracts, it is normal for the Employer to provide limited information in the Employer's Requirements, rather than the detailed documentation characteristic of a traditional bills of quantities project. It is these Employer's Requirements that may be changed in a process analogous to variations in a traditional project. To distinguish design and build from the traditional process, the term used is "Changes" rather than variations, but essentially the process is the same. The Employer may require changes (3.9) and "Change" is defined (5.1).

Valuing changes follows the same approach as with the traditional contract except that the Contract Sum Analysis is the pricing reference rather than bills of quantities. In Schedule 2, Part 1, Section 2 of the contract it is a *requirement* for the contractor, prior to doing the work, to quote for a change in the same way as a Schedule 2 Quotation in JCT/Q2016, with the same provisions for using measurement and valuation if the quotation is not agreed. The Contract Sum Analysis may or may not be helpful in valuing variations (less so if a simple elemental cost breakdown, more so if a schedule of rates or full bills of quantities) so the valuation of a change cannot be approached with the same confidence as with SBC/Q2016.

The Contractor's Proposals are not subject to variation, but the Employer may allow them to be varied, for example, to comply with building regulations, at no additional cost. It is also permissible for the Employer to effectively order a change to the proposals by changing the Employer's Requirements (these include the "addition, omission or substitution" of any work), but then this will require financial adjustment.

Two examples of possible modifications illustrate how this contract deals with changes:

1. Employer's Requirements make no mention of foundation design for a building, but it proves necessary for the contractor to modify the proposed design (shown on drawings in the Contractor's Proposals) as ground conditions are worse than expected. The Employer might allow the modification, but it will not be a change in Employer's Requirements and will not have a financial effect for the Employer.
2. Employer's requirements make no mention of the type of windows, but the Contractor's Proposals are for double glazed PVC units. On reflection, the Employer prefers sustainably sourced hardwood windows and instructs a change in the Employer's Requirements to allow for this. The cost of the change is paid by the Employer in accordance with (5.4).

As the design and build contract is not formally administered by an architect or other contract administrator, there are other changes of practice and terminology. For both interim and final payments, the contractor is given the primary duty to present financial information, with the Employer checking the amounts. Rather than an architect issuing a final certificate, the Contractor issues a "Final Statement" – the dispassionate term "certificate" being removed throughout the form. Should the Contractor not come up with the statement then the Employer may take the initiative by issuing an "Employer's Final Statement". Although neither architect nor contract administrator is mentioned by name, in many instances the issue of instructions and approval of statements will be scrutinised by a similar professional in the guise of the "Employer's Agent".

JCT MWD2016 contract provisions

The basic approach of SBC/Q2016 is maintained in a simplified form in this contract. The Architect is empowered to issue instructions in Clause 3.6 and to agree prices for variations prior to the work being carried out (3.6.2) in a simplified version of an SBC/Q2016 Schedule 2 Quotation. Failing agreement, the Architect is to value work on a fair and reasonable basis using relevant prices in whatever pricing document comes to hand – Contract Specification, Works Schedules, or Schedules of Rates. Simpler documents are likely to be of limited use and the Architect is likely to rely more on prior quotations or negotiating a fair price.

Clause 3.6.3 also rolls up considerations of indirect cost claims by indicating that the Architect shall include direct loss and/or expense incurred by the Contractor due to progress of the Works being affected by the variation. Thus, if for example, an instruction required that scaffolding be kept on site longer than expected, the Architect could include the cost here.

PRACTICE

SBC/Q2016

In SBC/Q2016, the Quantity Surveyor is responsible for preparing the final account (4.25.2.2) and this is a useful source of work for quantity surveying practices. There are two approaches, either drawing up an account towards the end of the contract period, with the final adjustments being made after practical completion, or measuring financial adjustments during the progress of the works as instructions are issued. The former procedure is older, but with an increased need for continuous projected cost information, the latter is now considered better practice.

Letting production of the final account lag construction has the advantage of being able to rationalise measurement and simplify the accounting process. For example, often several instructions are issued covering the same element of work. Hindsight allows these instructions to be grouped as a complete set under one item (e.g. "Variations to sub-structure"). Lagging the final account also allows the work to be fitted around more pressing matters in the office. Thus, final account production can be used as filling-in work. Measuring work as instructions are issued allows a timely financial contribution but is much more fragmented and can lead to abortive measurement when there are variations on variations.

Whatever method is used, it is usual to measure items in the account on an adds and omits basis, rather than attempt to value the net difference between new and original designs. The "omits" represent work omitted from the bills of quantities. They are cross-referenced to the bills (for auditing purposes) and kept as short as possible. The text of items rarely needs to be re-written, but sometimes "part quantities" need to be abstracted from original dimensions or measured on site. The "adds" represent the new or revised work. They are measured in much the same way as pre-contract taking-off, except that a larger proportion of work tends to be measured on site. However, measuring from drawings is often preferred as this represents the designer's correct intentions rather than the contractor's interpretation of those intentions (this is particularly important in relation to the depths and widths of foundations, which may exceed design requirements). Addition items are cross-referenced to the bills if possible for auditing purposes and they are valued initially at bill rates following the Valuation Rules in Clause 5.6.1.

Sections of the final account

These loosely follow the format of SBC/Q2016 Clause 4.3 and typically include:

PROVISIONAL SUMS

Each provisional sum is omitted and any work instructed in its place inserted in the same way as measured works additions. Provisional sums are divided into Provisional Sums for Defined Work or Provisional Sums for Undefined Work (NRM Part 2,

Rule 2.9.1.2). For the former, the work envisaged in the Provisional Sum is sufficiently defined for the Contractor to price for necessary preliminary items (perhaps a draft drawing of the work has been produced showing handling and approximate installation requirements). For the latter, a figure may need to be assessed to cover programming, planning and pricing preliminaries. This should be assessed as part of the Provisional Sum item and not as a claim[5] (e.g. under Clause 4.20).

APPROXIMATE QUANTITIES ACCOUNT

Approximate quantities items are omitted in their entirety and any such quantities instructed are added at the rates indicated against the approximate quantity in the bills.

MEASURED WORK ACCOUNT

The account is often presented in an order loosely following an elemental format. Items of account are generated either on a grouped or Instruction basis with omits first followed by adds.

DAYWORKS ACCOUNT

This will omit Provisional Sums for dayworks and add the total of agreed dayworks for Labour, Materials, Plant, Overheads and Profit, either at rates inserted in the Bills of Quantities (BQ) or at nationally agreed dayworks rates.

SCHEDULE 2 VARIATION ACCOUNT

These may also contain variations to the accepted variations quotation (variations on variations!). These variations are not valued under the normal valuation rules (5.6.1) but are based on a fair assessment (5.3.3).

SCHEDULE 2 ACCELERATION ACCOUNT

The Employer may request a quotation for accelerating the work to a new "Completion Date" – an agreed date overriding all extensions of time previously given (Schedule 2, 2.1). Once agreed, the quotation is added to the final account (4.3.2). If the quotation is rejected, the cost of preparing the quotation may be added to the final account (Schedule 2, 5.2).

CLAIMS

Agreed amounts for claims under Clauses 4.20–4.24 will be detailed separately. The subject matter of claims is covered in Chapter 6. If amounts for loss and expense related to a variation are agreed in advance (under the provisions for a contractor's Schedule 2 Variation Quotation), they would be included with the Schedule 2 Variations Account.

FLUCTUATIONS

Any increased costs under the fluctuations options, A B or C will be detailed. The subject matter of fluctuations is covered in Chapter 11.

OTHER ITEMS

Other exceptional items could include reductions in value where defective work has been accepted, where inaccurate setting out is allowed etc.

DB2016

As the organisation producing the final account is, in the first instance, the contractor rather than a consultant quantity surveyor, there is less incentive to be analytical in presenting the detailed reconciliation envisaged with SBC/Q2016. However, where the account is prepared by a professional quantity surveyor working for the contractor, practice is likely to be very similar. The Employer's Agent may demand full information before payment and, should disputed items be included, is more likely to use the provisions incorporated in the contract for paying less subject to further negotiation or adjudication.

JCT MWD2016

This contract puts the onus on the Architect to present the final account, but this is to be based on the receipt of "all documentation reasonably required for computation of the final payment". The Architect could argue that full documentation has not been provided and this would effectively transfer the primary duty to the Contractor. In practice, preparation of the final account is likely to be done by the Contractor in any event, with the Architect checking the amounts and issuing the final certificate. Details of the account should follow practices outlined above, with each major section separately identified to allow and audit trail from item to authorisation under the contract.

SOURCES OF INFORMATION FOR THE ACCOUNT

For variations and re-measurements related to sub-structure or subsequently covered up, it is important to have accurate information before work is hidden. The sources of information relied on might include:

1. Drawings – These are often the best guide as they represent what should be built provided an instruction has not relaxed a drawn requirement.
2. Site measurements – These are usually needed to determine the depths of excavations, inspection chambers etc. The widths and plan areas are usually taken from a drawing. Site measurements may be performed by:
   (a) A quantity surveyor, architect or employer's agent on site
   (b) Clerks of works or resident engineers
   (c) Contractor's surveyors or agents
   (d) By inspection of photographs

FINANCIAL STATEMENTS

As part of the regular interim valuation process, it is usual for the relevant professional, depending on the contract in use (quantity surveyor, contractor, employer's agent or architect) to give an indication of the projected final account in the form of a financial statement. This requires that the account be kept reasonably up to date. Appendix 1 is a typical format.

APPENDIX 1 CONTRACT FINANCIAL STATEMENT

Contract name and number	£	£
Contract sum		
Less contingencies		
Adjustment of provisional sums		
Adjustment of approximate quantities		
Measured variations to date		
Measured variations anticipated		
Schedule 2 Variations Account		

Schedule 2 Acceleration Account		
Claims agreed		
Claims anticipated		
Fluctuations to date		
Fluctuations anticipated		
Exceptional items		
Anticipated total cost		
Approved total cost*		
Net extra/Saving		

* Depending on the financial protocol in place, this may be more than the contract sum.

APPENDIX 2 EXAMPLE FINAL ACCOUNT ITEM

Measured work						
Item no 67						
Omit vinyl flooring to kitchen and softwood skirting						
Add quarry tile paving and tiled skirting						
AI 15(6)						
			Omissions			
		B/Q		Qty.	Unit	Rate	£
	14.00	83/D	4 mm vinyl flooring			
	12.00					

	22.50						
	9.00						
	15.00						
	6.00			461	m^2	14.00	6,454.00
	157.00	75/A	150 mm skirting	157	M	3.00	471.00
		127/G	Paint skirting	157	M	1.50	235.50
			Total omits				**7,160.50**
			Additions				
			12 mm quarry tile paving	461	m^2	19.50	8,989.50
			Coved quarry tile skirting	157	M	9.00	1,413.00
			Total adds				**10,402.50**

NOTES

1. *Romeo and Juliet*, William Shakespeare.
2. The Public Contracts Regulations at http://www.legislation.gov.uk/uksi/2015/102/introduction.
3. Costs might be particularly high if the quotation involved producing design information, but the Contractor should gain agreement to payment prior to starting design.
4. See Chapter 6.
5. See Chapter 6.

5

Progress

INTRODUCTION

In the absence of express terms covering project timing, a contractor would have a reasonable time to complete work. This is much too vague for all but the smallest contracts and the JCT contracts, in common with most other standard forms, stipulate how progress is to be maintained. They also spell out the remedies available for delays, whether caused by the parties themselves or some outside influence. It is also necessary to identify when the works are actually complete, so that contractual remedies for delay can be instigated. The JCT contracts SBC/Q2016, DB2016 and MWD2016 adopt a fairly common terminology, much of which has been in use for a considerable time and is familiar to the construction industry.

TERMINOLOGY

The contracts work by stipulating key start and finish dates which may be modified as the result of extensions of time. The works may actually be finished before the original date, or date as extended, or could be late, giving rise to a liability to liquidated damages. Recording the actual date of finishing the practical work on site (Practical Completion) is, therefore, very important.

Date of possession

This is the date on which the Contractor will be given possession of the site in order to start the works. It is entered in the Contract Particulars prior to signing the contract. On possession, use of the site by the Employer is restricted to express provisions in the contract. In MWD2016, the date of possession is named the Works Commencement Date.

THE CONTRACT TIME LINE

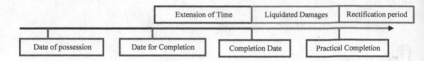

Figure 5.1 The contract time line

Date for completion

This is the original intended dated that the parties envisaged that the contract works would be finished. It is entered in the Contract Particulars prior to signing the contract.

Completion date

This is the revised date that the parties may later contemplate after the operation of the provisions in the contract for extending time, or accelerating the works. If neither occurs, then the calendar Date for Completion becomes the Completion Date. In MWD2016, the term remains Date for Completion but is supplemented by the phrase "or any later Date for Completion fixed under clause 2·8" (the extension of time clause).

Practical completion

This is when the works can actually be taken to be complete. The point of "Completion" is not defined by the contract and building works usually have a very long tail of minor snagging items extending to what could be termed maintenance. However, a practical approach might indicate that it is achieved when the finished works can be used as intended. Some minor works of tidying up and finishing may remain.

Partial possession

It is possible for Practical Completion to be achieved for parts of the works. This is by agreement after the works have started, but it must be possible to split off suitable self-contained parts of the project, as the contractor will need unimpeded access to the rest. MWD2016 does not make express provision for partial possession.

Sectional completion

It is also possible to agree <u>in advance</u> that the works will be completed and handed over in sections. In SBC/Q2016, this is done by the parties mentioning sections in the Sixth Recital (Fifth Recital in DB2016) and completing in the Contract Particulars the relevant dates of possession and completion of each section The sections should also be identified in the Contract Drawings and Contract Bills (The Employer's Requirements for DB2016). The wording of the contract allows for progress pinch points for either the Works as a whole or sections, where relevant. As with Partial Possession, for Sectional Completion to work, it must be possible to divide the project into clearly defined parts, as the contractor will need unimpeded access to sections still in its possession. MWD2016 does not make provision for sectional completion, and if it is required, it would need to be written into the form as an amendment.

Early use

It is possible for the Employer to agree with the Contractor early use of all or part of the Works, without formal possession being handed over. It is, however, important to ensure that insurers are notified (by the Contractor or Employer as appropriate – see chapter 10 on insurance), that the insurers accept the early use and that additional premiums for the use are paid if required. In addition, the Employer will not be covered under Works insurance (Schedule 3 A or B) for any contents brought in, or acts of employees affecting the project and must arrange appropriate insurance for this. Operationally, it is also important that the mixed use of the project can be managed safely by Contractor and Employer. MWD2016 does not make express provision for early use.

Extension of time

Additional time may be given for the Contractor to reach the Completion Date for various reasons either resulting from actions by the Employer or its agents, or from some events wholly beyond the Contractor's control.

Liquidated damages

If the Contractor has not finished by the Completion Date, liquidated damages may be exacted from the Contractor by the Employer. Liquidated damages are meant to be a genuine pre-estimate of expected loss due to lateness and are therefore normally at an amount set in the Contract Particulars as a sum per unit time of delay.

Rectification period

Once Practical Completion has been achieved, the Contractor is given the opportunity to put right defects for a set time (6 months in SBC/Q2016 and DB2016, 3 months in MWD2016). Thereafter, the Employer may use others to put right defects and take action against the Contractor if appropriate.

Final adjustment

In SBC/Q2016, the Contractor has 6 months from Practical Completion to provide the Architect or Quantity Surveyor with all "documents necessary for the adjustment of the contract sum". Thereafter, the Architect/Quantity Surveyor has 3 months to produce the final account ("statement showing all adjustments" to the contract sum – 4.25). In DB2016, the Contractor has 3 months from practical completion to produce a "final statement". If not forthcoming, the Employer may encourage its appearance by giving notice that, unless it is forthcoming within 2 months, the Employer will produce its own statement (the "Employer's Final Statement"). In MWD2016, there is no separate final adjustment period. The Contractor simply has 3 months to provide the supporting figures and the Architect up to 33 days thereafter to issue the Final Certificate.

Acceleration quotation (and consequent revised completion date)

With both SBC/Q2016 and DB2016, the Employer may request a quotation for accelerating the work. The quotation, if accepted, substitutes a new completion date for the then current completion date (the date currently in prospect for completing the work allowing for extensions of time). There are no express provisions for acceleration in MWD2016.

HOW THE CONTRACTS OPERATE

Topic	SBC/Q2016	DB2016	MWD2016
Fundamental obligation A major contract term requiring the Contractor to carry out and complete the works. Breach of this term goes to the heart of the contract and would lead to Termination of the Contractor's employment under Section 8	2.1	2.1	2.1
Starting date The Contractor will be given possession of the site (or a Section) on the date mentioned in the Contract Particulars. The Contractor will get on and finish by the Completion Date (Date for Completion as amended in MWD2016)	2.4	2.3	2.3

Deferred starting date Possession may be deferred by the Employer for up to 6 weeks, provided a note to this effect is included in the Contract Particulars, but the Contractor can claim an extension of time as a result (but not loss/expense)	2.5	2.4	–
Postponement The Architect (SBC/Q2016) or Employer (DB2016) can postpone any work. Note, the Contractor could get an extension of time and loss/expense as a result of postponement. Postponement, as distinct from deferment, relates to any work during and before starting on site	3.15	3.10	–
Practical Completion The relevant clauses set out the procedures adopted when the works (or a Section) are complete, subject to final rectification of defects. The sequence is as follows: Architect issues Practical Completion Certificate (SBC/Q2016, MWD2016), or Employer issues Practical Completion Statement (DB2016) when he/she thinks it has been achieved This has the effect of: • Starting the Rectification Period • Ending the Contractor's responsibility for insuring the Works • Ending the Contractor's liability for liquidated damages • Halving Retention held • Starting the period for final adjustment of the contract sum	2.30	2.27	2.10
Liquidated damages In essence, the Employer may deduct liquidated damages at the rate set out in the Contract Particulars, provided that the	2.32	2.29	2.9

Architect certifies that the Contractor has failed to complete by the Completion Date and that the Employer has given notice of an impending deduction before the date of the Final Certificate. There is a provision for repayment if an extension of time is later granted. In DB2016, provisions are similar, but the Employer issues a "Non Completion Notice" if works are not complete and a notice of an impending deduction before the due date for final payment MWD2016 is simpler but follows the same form as SBC/Q2016. There is no requirement for a certificate – if the works are not complete by the Date for Completion as extended, the Contractor pays liquidated damages until the works are complete. This is provided that the Employer has given notice of an impending deduction before the date of the Final Certificate In all forms, deduction of liquidated damages is a "set off" by the Employer and is subject to the payment notice requirements of the contract and the Construction Act – i.e. a Pay Less notice issued at least 5 days before the final date of payment of an interim or final certificate (see also Chapter 14)			
Delay notice The Contractor is required to give written notice to the Architect (SBC/Q2016), or Employer (DB2016) of likely or actual delays, whether "Relevant" or not, but also specifying causes and identifying what it considers to be any delays caused by Relevant Events (Relevant Events are events that could qualify for an extension of time). The Contractor is required to specify in the notice, or as soon as possible thereafter, the expected effects of the occurrence of the events and estimate the likely length of delay	2.27	2.24	2.8

beyond the Completion Date. The Contractor is required to update notices as necessary MWD2016 simply requires that the Contractor gives written notice to the Architect that the works will not be completed by the Date for Completion			
Fixing Completion Date (Extension of Time) The Architect (SBC/Q2016), or Employer (DB2016), is required to:	2.28	2.25	2.8
• Give an extension of time by fixing a later Completion Date if the events specified in the Contractor's notice are "Relevant" and that the completion of the works is likely to be delayed beyond the Completion Date • Notify the Contractor in writing of the decision stating the extension granted for each Relevant Event and for each Relevant Omission (the extent to which an extension of time has been reduced by taking into account omissions of work under the Variations (SBC/Q2016), or Change (DB2016) clause since the last extension of time). Omission of work is the only way that the Extension of Time can be reduced • Carry out this duty within 12 weeks of receiving the Contractor's notice or before the Completion Date (if possible), whichever is the earlier MWD2016 is much simpler; the Architect simply gives a reasonable extension of time and notifies the Contractor			
Reducing an extension of time The Architect (SBC/Q2016), or Employer (DB2016), may fix an earlier Completion Date if work is omitted (A Relevant Omission), but not earlier than the date fixed at the last Extension of Time	2.28.4	2.25.4	–

Review of extensions of time The Architect (SBC/Q2016), or Employer (DB2016), may review all extensions of time after the completion date and up to 12 weeks after Practical Completion. He/She can increase them but can only reduce them if work is omitted since the last Extension of Time	2.28.5	2.25.5	–
Provisos • Contractor's duty of mitigation (to try and minimise delays) • No reduction of original time (not earlier than the Date for Completion – the original date inserted in the Contract Particulars) • Does not apply to Pre-agreed Adjustments – extensions of time agreed at the time of accepting a "Schedule 2" variation quotation (SBC/Q2016), or a "Schedule 2, part 1, section 4" contractor's change estimate (DB2016) – unless work subject to that quotation is omitted since the last Extension of Time	2.28.6	2.25.6	–
Relevant Events For SBC/Q2016 and DB2016, there is a detailed and comprehensive list of Relevant Events (Reasons for an Extension of Time) and, although not reproduced here, they are self-explanatory. They are of two broad types: 1 Events related to actions by the Employer or the Employer's agents. These include late issue of Architects (SBC/Q2016), or Employers (DB2016) instructions, failure to give possession of the site and work by "Employer's Persons" – parties directly contracted to the Employer	2.29	2.26	2.8

2 Events wholly outside the control of either party. These include "force majeure", "exceptionally adverse weather conditions" and insured events (fire, flood etc.) – called "Specified Perils" in these contracts Events are not spelled out in MWD2016, they simply have to be beyond the control of and not caused by the Contractor. Compliance with Architect's instructions is specifically drawn out as being relevant. Actions of suppliers and subcontractors are specifically drawn out as not being relevant			
Early use The Employer may agree with the Contractor to have use of the site before practical completion of the Works or a Section. Insurers are to be informed to ensure that the use will not prejudice the cover. For Option A of Schedule 3, where the Contractor is insuring the works, any additional costs for insurance are to be added to the contract sum	2.6	2.5	–
Partial possession by Employer By agreement between the parties, the Employer may take possession of part of the Works before the whole project has reached Practical Completion. In such a case, the Works are split in value terms so that: • Most of the Practical Completion provisions above for the Relevant Part that is handed over are implemented – for example insurance and retention • The liquidated damages liability is proportionately reduced	2.33–37	2.30–34	–

Acceleration quotation		
The Employer may request from the Contractor proposals for accelerating the work and making an appropriate agreed payment. The Contractor may decline the opportunity but is expected to give reasons. Alternatively, an acceleration quotation is required to be submitted within 21 days and accepted/rejected by the Employer within 7 days On acceptance, the Completion Date is altered accordingly (may be earlier than the Date for Completion inserted in the Contract Particulars) and the cost is added to the Contract Sum. The new Completion Date becomes the reference point for later extensions of time, payment of liquidated damages etc. Should the quote be bona fide, but not accepted, the cost of preparing it may be added to the contract sum	Schedule 2, part 2	Schedule 2, part 2, section 4	–

6

Claims

INTRODUCTION

The term "claim" is not one that is used by the JCT standard forms of contract, but general usage indicates that it is a term relating to the request for additional payment by a contractor or subcontractor for work executed beyond the normal provisions of the contract in such a way that the normal provisions do not provide adequate reimbursement.

But what are the normal provisions of a contract? One definition might be those provisions whose ground rules are determined by written conditions of contract as opposed to those whose ground rules rely on implied terms, outside the written conditions, or those based on a moral, rather than legal obligation.

Claims within express terms

Modern standard forms of contract attempt to provide comprehensive and complete sets of express rules to deal as far as possible with all eventualities and all demands for extra payment. This avoids uncertainty and expensive recourse to litigation to resolve "extras". Features common in these rules are:

- A clearly defined procedure for processing claims.
- A provision for valuing claims, tied back where possible to the original contract sum, for example by using rates inserted in a bill of quantities, contract sum analysis or schedules of rates.
- A provision for impartially certifying extra time and money.
- A procedure for settling disputes within the contract framework (adjudication/arbitration).

Claims based on implied terms

Redress is occasionally sought outside the provisions of the contract, relying on implied terms. In some cases, where there may have been a breach by the employer of a major term, the contractor may seek to have the contract set aside (in the courts by **rescission**) and payment for any work ordered based on a reasonable assessment rather than related to contract rates. The attraction of this course of action is amplified when the contract is making a loss based on contract rates. A reasonable assessment (often termed "quantum meruit") would allow the contractor to recover costs, whereas payment on the basis of contract rates would compound the loss.

Claims based on a moral obligation (ex-gratia claims)

An ex-gratia claim is not based on a legal obligation to pay, but on an appeal to fairness, morality or longer term cooperation. It is a non-contractual gift by the client in the spirit of a consolation prize or a contribution to a contractor's bad luck. It is sometimes agreed in circumstances where to refuse might cause a larger and more expensive calamity, such as the liquidation of the contractor or loss of goodwill from an otherwise satisfactory firm. Payments in response to such claims are not supported by consideration and will not normally be enforced in the courts.

NORMAL PROVISIONS WITHIN JCT CONTRACTS

The normal rules for dealing with additional payments are contained in such clauses as SBC/Q2016, 5.6 – the Valuation Rules – and provide in the main for payment on a quantity-related basis. These rules also cover the effect of, for example, an omission of work increasing the cost of any remaining work (5.9) but expressly exclude costs caused by disturbance of the regular progress of the works (5.10.2).

By and large, the rules will adequately reimburse a contractor for additional payments for variations, but costs not related to quantities may be incurred. With the increased mechanisation of building work and pre-fabrication of components, fewer payments are based on quantity and more are related to logistics and duration. Causes driving these costs are related to factors such as late issue of Architects' Instructions, delay by Employer's Persons, Architects Instructions increasing the quantity of the work and thus the time it takes to complete the work etc.

It is not necessary for there to have been a delay on a contract for additional costs to have been incurred, but by far the most common claims are in association with delays. Thus, clauses SBC/Q2016 2.28 (Fixing Completion Date) and SBC/Q2016 4.20 (Loss and Expense) are often termed the "Claims" clauses. There are corresponding clauses in DB2016 and, in a simplified form, in MWD2016.

Additional payments not related to quantities include those for:

- Delay
- Uneconomic working
- Loss of profit and interest charges

Delay costs

Relief of liquidated damages – an extension of time

This is usually the first item that the contractor will pursue and the causes ranking for relief are listed (for SBC/Q2016 in 2.29). Many of these causes relate to the contractor being prevented (by the Employer or its agents) from carrying out work to the agreed timescale. However, it need not necessarily be the employer's fault that delay has occurred. For factors completely outside the contractor's control, such as the occurrence of exceptionally adverse weather, an extension may be claimed. The factors also include the occurrence of "specified perils" such as a fire delaying the work even though the Contractor may have caused it! It is the policy of some forms of contract to allow extensions of time for such factors. This avoids contractors having to price for the risk of the occurrence and, therefore, reduces the general levels of tenders for building work. It is also possible for an Employer to insure some risks of having to give an extension of time and lose liquidated damages in the event of delay.

Additional on-site overheads

If the cause of a delay is the employer's or the employer's agents, this is usually the first and only positive cost item a contractor will attempt to recover. Additional costs will relate to time-based charges often published as preliminaries in Bills of Quantities, Contract Sum Analyses, or Schedules of Rates and thus fairly readily ascertained. Details include extended hire charges for scaffolding, site cabins, cranes etc. Detailed records of tender breakdown and actual costs are important supportive evidence for sustaining a claim.

Additional general off-site overheads

These costs relate to the "general overhead" incurred in running a contracting business (example items could be senior management, head office rent and company cars). These overheads are insufficiently specific to be priced to a particular project and will usually be assessed by calculating the total overhead for the past year and expressing it as a percentage of the net turnover of the company. A lump sum, based on the percentage, will then be added to the tender. If the duration of the project is extended, particularly without sufficient notice to reorganise activities, then the overhead allowance will be stretched over a longer period, reducing the percentage recovery.

If substantial enough to be worth losing goodwill over, "Hudson's formula" may be tried:

$$\frac{(\text{Number of claimed weeks} \times \text{Head office overhead content})}{\text{Contract period}}$$

This formula (named from a standard legal work on building contracts[1]) assumes that overhead is being accrued on a strictly time basis and that extra time will need proportionately the same overhead. However, this will not be correct if:

i. Delay relates to insignificant volumes of physical work (e.g. near the end of a project, when most work is complete).
ii. Resources are released by the delay, for example, by re-deploying staff and equipment to other projects. This will allow the overall turnover to be maintained and not reduce the percentage recovery.

For these reasons, claiming general off-site overheads (as with claiming loss of profit – see below) is difficult and contentious.

Increased costs on work pushed back (fixed priced contracts only)

For fixed price contracts, contractors take the risk of inflation increasing factor costs (labour, materials, plant and overheads) and will add an amount to cover the risk. When work is delayed, factor costs may increase further than envisaged in pricing. Increased costs may apply to all remaining work after the Date for Completion, but the real cost may be very much higher (e.g. the bulk of the work may still be executed within the contract period but at a much later date). However, proving and justifying the extra amounts will be difficult unless extremely good records are maintained.

Seasonal effect

A delay may mean that remaining work is now to be executed in the winter rather than the summer. It is unlikely that a benefit could be obtained by an employer from the opposite effect (work being pushed from a winter period into a summer period) as the instigating action will have come from the employer. Winter working involves shorter periods of daylight and a greater incidence of adverse weather, which reduce the productivity of both labour and plant.

Uneconomic working

Uneconomic working of labour

Whether or not delay has occurred, an Architect's or Employer's Instruction, for example, issued very late may have caused a disruptive effect on the planned work pattern of the contractor. Delays might be avoided by more intensive working involving overtime, or by more intensive supervision, both of which increase costs. By far the biggest problem for the contractor is proving the disruption. It is necessary to separate costs as actually incurred from notional costs (costs as would have been incurred had the event not happened). A comparison between actual cost of labour and planned cost of labour in the tender is not sufficient because of possible inefficiencies of the contractor due to other concurrent causes.

The only satisfactory action a contractor can really take is to keep extremely good records of labour and plant costs as incurred and forewarn professional advisers of the likely cost of instructions etc. as soon as they are issued (the latter is also a requirement of SBC/Q2016).

Uneconomic ordering of materials

A contractor will normally plan activities for a project well in advance of construction and arrange for efficient ordering of materials. This is important as many heavy materials such as bricks, cement and aggregates are much cheaper to transport in bulk loads at maximum vehicle quantities. Additional costs may be incurred, where, for example, variations are ordered using the same materials, but with different delivery times and quantities. Such additional costs may include:

1. Cost of inefficiently using plant and vehicles (e.g. digging small quantities, or only partly filling lorries).
2. Cost of transporting additional quantities separately from a bulk order.
3. Cost of handling varied quantities on site and in the yard.
4. Additional head office expenses in dealing with varied material requirements.
5. Additional site office expenses in dealing with varied material requirements.

Uneconomic use of plant

Where items of plant are priced in unit rates rather than as preliminaries, additional quantities or late instructions may mean extra transportation charges, use of smaller uneconomic plant or hand working instead of machine etc. Again, accurate records are essential, but this may be easier to prove than uneconomic use of labour.

Loss of profit and interest charges

Loss of profit can be a contentious claim as simply not doing planned work does not stop working elsewhere! There would need to be idle time, or delay

Claims

inhibiting use of resources on alternative sites. The claim also depends on work being available at similar profit rates elsewhere. A common circumstance for loss of profit claims is where large omissions of work are made at the last moment, after resources have been geared up and operatives/plant are left standing around. Loss of profit is a valid item if the Employer should have been able to envisage its loss in causing the claim. Exceptional loss of profit (e.g. where a delay stopped the contractor working on a particularly lucrative contract the Employer was unaware of) could not be claimed.

Claims for interest are rarely pursued against an employer, unless the dispute is taken as far as the courts. In principle, interest payments can be claimed from rejection of the claim up to commencement of any Arbitration or Court action. The Late Payment of Commercial Debts (Interest) Act 1998 allows simple interest to be claimed on late payments. (Provisions based on this Act are included in SBC/Q2016 in 4.11.6.)

Re-rating or claim?

There is provision in the normal rules for valuing variations to make adjustments to rates for carrying out remaining work if conditions change as the result of a variation. The changed conditions are, effectively, treated also as a variation and are subject to new rates (perhaps based on a fair valuation rather than contract rates). This changing of rates to suit new conditions is informally termed "re-rating". Although often subject to considerable scrutiny and negotiation, re-rating is not as contentious as pursuing a claim based on disturbance of regular progress and it is sometimes possible to reclassify claims as quantity-based payments.

Several of the above have elements of cost, which could be related to quantities and one simple claims tactic to reduce the volume of contentious items is to extract as many quantity-related heads of claim and to treat them as requests for re-rating under the valuation of variations rules. These can then be accepted or rejected in principle and, if accepted, negotiated on the basis of objective price data.

HOW THE CONTRACTS OPERATE

Extensions of time have been dealt with in the previous section, so this section deals only with disturbance payments. The contracts require a proactive approach from the Contractor in forewarning of likely expenses, therefore giving the Employer or Architect the opportunity to take mitigating action.

Topic	SBC/Q2016	DB2016	MWD2016
Procedure The Contractor notifies the Architect of the effect of a Relevant Matter on progress and expense. Expense must arise as a result of deferment of giving possession of the Works, or disturbance of the regular progress of the Works For MWD2016, direct loss and expense is allowed for only in relation to the regular progress of the Works being affected by compliance with a variation instruction and the cost is simply to be included in the Architect's valuation of the variation. However, "Variation" is framed more widely in this contract to include any "other change in the Works or the order or period in which they are to be carried out" and is likely to cover most Employer driven events giving rise to claims	4.21	4.19	3.6.3
Within 28 days, the Architect, or Quantity Surveyor, is required to notify the Contractor of the ascertained amount of the loss and expense. As this ascertainment could clearly differ from the Contractor's assessment, the Architect/QS must give enough detail for the Contractor to "spot the difference" – presumably as an aid to further negotiation. The ascertained amount is added to the Contract Sum (4.23) and paid on the next Interim Certificate (4.14.2.3)	4.21.4		–
For DB2016, the Contractor notifies the Employer and the loss and expense is ascertained in the same way by the Employer. The task itself is likely to be carried out by the Employer's Agent, if any		4.20.4	
Provisos – the Contractor is required to: • Make the application as soon as possible • Submit supporting details	4.21.1–3	4.20.1–3	–

Other routes to redress The clause is stated to be without prejudice to other remedies that the Contractor may possess	4.24	4.23	–
Relevant matters The matters for which the Contractor can claim Loss and Expense are detailed but can be summarised as those for which an Extension of Time would be given and for which the Employer is responsible	4.22	4.21	–

It should be noted that, although the extension of time and loss and expense clauses may appear to be linked (and they often operate together as "Claims Clauses" in practice), they are separate. The reason that they are linked is that delay is a common cause of loss and expense, not covered by other clauses in the contract. However, it is possible for a contractor to be involved in loss and expense without being delayed.

The principles of Extensions of Time and Loss and Expense can be summarised as follows:

- Should the event/matter be within the control of the contractor, there is no claim.
- Should the event/matter be caused by the Employer, there may be a claim for extra time and money.
- Should the event/matter be wholly beyond the control of either party, then the difference is split – an extension of time may be granted, but no prolongation expenses.

Negotiating claims

The legal principles and main heads of claims set out above provide a framework for their resolution, but whether an extension of time is granted or a loss and expense payment made depends (as far as reasonably possible) on matters of objectively determined fact. Agreeing the facts of the issue in question, not determining the framework, forms the basis for most claims negotiations. Examples of important factual matters include:

- Whether a relevant event or matter has occurred.
- Whether an event or matter has caused a loss.
- The extent of any loss.

Whether an event or matter has occurred

An example illustrating the difficulty in determining whether an event has occurred is related to the common request for an extension of time for "exceptionally adverse weather" (SBC/Q2016 – 2.29.9, DB2016 – 2.26.8). Whether the weather is normal or whether the weather is exceptionally adverse depends on time of year and location of site. It is not normally sufficient for a contractor to simply prove cause and effect – bad weather has stopped work! This is because, objectively viewed, another (perhaps more efficient or well equipped) contractor might not be affected in identical circumstances. A decision will, rather, usually be based on statistical frequency of occurrence – for example the event in question has only occurred at a frequency of once every 10 years.

A similar problem might arise where information is provided to a contractor in stages to suit the timetable of the project as evidenced by the Date of Possession and Date for Completion inserted in the Contract Particulars. A question might be whether information has been provided on time if it does not meet a more ambitious (but not contractually binding) programme produced by the contractor. An architect is not obliged to provide information to meet the contractor's programme but must meet the dates in the contract.

Whether an event or matter has caused a loss

A project may not be delayed overall if the work affected is not critical to progress. In any building work, many sets of tasks are carried out in parallel. Organising parallel working allows the duration to be shortened and gives rise to the concept of the "critical path" – that set of sequential tasks which takes longest and thus gives the project its overall duration. An event delaying or lengthening a non-critical task will not affect overall duration until it is so extensive that the set of tasks in which it occurs itself becomes critical. Until this happens, any delay will not affect progress and any delay expenses will be irrelevant.

Similarly, many relevant delays may be concurrent with other irrelevant events caused by the contractor or parties for whom it is responsible (concurrent delay). The question becomes one of which event has caused a delay. In some instances, the answer might be the predominant event or the first event to occur and it is a question of fact as to whether this event is or is not relevant. Similar problems arise with loss and expense payments where additional expense might be caused by a "relevant matter" or might be related to inefficiencies of the contractor unrelated to the matter (concurrent expense).

The extent of any loss

Evaluating loss for delay and disruption can be extremely difficult. For example, losses based on extended general overheads if determined in detail would require the assessment of portions of managerial time, secretarial support, company vehicles, head office rates and rent etc. attributable to a particular project at a particular moment in time. It is for this reason that formula reimbursement (Hudson's formula or similar) is used to assess general overhead claims and has been endorsed by courts. Similar problems arise with uneconomic use of labour, materials and plant; although assessment is possible if good records are maintained.

For site overheads, assessment is often easier, being based on schedules of identifiable items. However, even for site overheads, detailed assessment is necessary to determine (1) what items were on site at the time of the event, and (2) what the actual cost of the items were. Early in the project, site overheads might be extensive and disruption expensive; towards the end of the project (perhaps in final finishing and landscaping stages), there may be relatively few items remaining on site.

Claims and subcontractors

Events leading to claims between contractor and employer often originate with subcontractors as they are the parties on the ground directly affected by disruptive action. It is normal for the contractor to act as an agent, pursuing the claim on the subcontractor's behalf even though, contractually, the matter is between the two parties. Not all such events are, however, related to actions of the employer or its agents and these remain to be settled between subcontractor and contractor. Two common problems, not directly affecting the employer, are attendance and knock-on claims.

Attendance claims

It is usual for contractors to attend on subcontractors by providing either general items, (such as scaffolding, use of accommodation or cranes), or by providing special attendance – for example the provision of a temporary road for a concrete piling subcontractor. There may be disputes concerning the agreement to provide general or special attendance and settling these claims should not affect the employer.

Knock-on claims

Often one subcontractor will be affected by the actions of another, for example a delay by one subcontractor will delay the start of all the following subcontractors. Or, defective work by one subcontractor in a sequence of tasks might mean that subsequent work, carried out by other firms, will have to be dismantled at additional cost and with considerable delay. Unless the cause of the instigating event is related to the employer, these matters will not surface at the main contract level and will have to be settled between the contractor and each subcontractor concerned.

NOTE

1. Atkin Chambers (2019) Hudson's Building and Engineering Contracts, 14th edition, Sweet and Maxwell, London, UK, ISBN: 9780414073883.

7

Termination and insolvency

INTRODUCTION

Occasionally during the progress of a building project, things go very seriously wrong, with either party substantially not performing their part of the bargain. This might involve the contractor in completely failing to carry out any work, or working so badly or slowly that it becomes clear that no contractual targets will be approached. From the employer's side, failure to pay agreed interim amounts, not allowing access to the site, or physically impeding progress might have the same effect.

Although poor performance or lack of goodwill might sometimes cause these problems, they are often driven by the actual or impending insolvency of one of the parties. Insolvency leads to poor performance and termination is the result. The two concepts are, therefore, considered together in this section. The position of fundamental breach of contract and termination is considered first, with a separate consideration of insolvency and bankruptcy at the end of the section.

A slightly different problem, but with similar consequences, is the occurrence of an intervening event making the performance of the contract impossible. This is uncommon in building contracts as most events (such as destruction by fire) can be overcome and work can, eventually, continue. However, a building being refurbished might conceivably be irreplaceable and rebuilding could destroy its essence.

As with other contractual events, the JCT contracts endeavour to avoid recourse to general law and the courts on the occurrence of major failure in performance. The overall objects are to provide clear procedures and impartial decisions, relate associated costs to the original bargain wherever possible and provide a clear dispute resolution mechanism if needed. Achieving these objects introduces certainty and reduces costs for both parties.

TERMINATION

The general legal position

For a fundamental breach of contract by either party to a building contract, the other party may be able to treat the contract as having been repudiated and seek rescission in the courts.

The breach must be of a major term, going to the heart of the contract, and not be a mere breach of a warranty. In building contracts, partial performance of the contract in such areas as late completion, normal defects, late payment (within bounds) and variations that retain the basic character of the project would not amount to a fundamental breach. Remedies for such actions (or inactions) would lie in damages based on the contract.

The main heads of fundamental breach in building contracts hinge around total non-performance by the contractor and total non-payment of an amount certified by the employer. The non-performance must not be merely late performance (unless involving persistently repeated late performance) and the non-payment must be total, or very late. As interim payments are, however, usual in building contracts, non-payment of an interim payment might be a valid ground for rescission. The general intent shown by the defendant would be an unwillingness to be bound by the terms of the contract.

Frustration

In addition to fundamental breach of the contract by either party, it may become incapable of performance because of "frustration". An intervening event, completely beyond the control of the parties, may mean that the contract is frustrated and cannot be performed at all. Thus, the destruction of the irreplaceable building mentioned above might frustrate the performance of a building refurbishment. The event relied on must render the contract completely incapable of performance and not merely more onerous. The doctrine of frustration is used sparingly and was often associated with sinking ships (where the subject matter of the contract had been lost) and, in building contracts, with declarations of war, when building was banned by statute.

Quantum meruit

In the situation where a contract has been frustrated, the common law position regarding damages is that the loss "lies where it falls". This means that neither party gains any benefit from the contract. The equitable remedy of quantum meruit does however allow an aggrieved party to recover payments from the other party to the extent that they are "merited".

A building contractor could therefore (should the contract be frustrated) recover some payment based on the cost of labour, materials, plant and overheads actually incurred. This corresponds to cost recovery on the basis of dayworks rather than contract rates.

Much more importantly, in the event of a fundamental breach of a building contract by the employer, the contractor may be able to claim payment after rescission on the basis of quantum meruit as the contract would have been set aside and contract rates would not directly apply. This would have the effect of converting (possibly) loss-making contract rates to cost-recovering daywork rates. In many of the more acrimonious building contract law cases, contractors have claimed fundamental breach by the employer and for payment on a quantum meruit basis!

Termination – standard form of contract provisions

Following the philosophy of the forms in being as comprehensive as possible, there are extensive clauses covering the situation of fundamental breach of contract by either party and of what might otherwise be classed as frustrating events. The philosophy goes as far as the contracts referring to the termination of the **employment of the contractor** rather than the **termination of the contract** – the contract provisions remain in force, whilst employment is terminated and the after-effects are dealt with.

How the contracts operate

Topic	SBC/Q2016	DB2016	MWD2016
Default by contractor These clauses allow the employer to terminate the employment of the contractor under the contract. This is for events generally classed as being breaches by the contractor including persistently poor progress, persistent failure to rectify defects, suspension of the works without good cause, commission of an act of insolvency, or an act of corruption	8.4–6	8.4–6	6.4–6
Default by employer These clauses allow the contractor to terminate its own employment for events classed as being breaches by the employer – failure to pay, failure to produce necessary drawings and the like, failure to allow access to the site, commission of an act of insolvency	8.9–10	8.9–10	6.8–9

External events These clauses allow either the employer or the contractor to terminate the contractor's employment in the event of major acts occurring to obstruct the performance of the project. These would correspond to frustrating events but are wider. To insert an element of objectivity to a notice of termination, the triggering cause must cause a substantial suspension of the works (by default for two months). As an event might be caused by the contractor (e.g. a major fire destroying the works), the contractor is not allowed to terminate its own employment if it has been negligent For public contracts the employer may also be able to terminate the employment of the contractor where there has been "substantial modification which would have required a new procurement procedure" and for serious infringement of EU procurement rules[1]	8.11	8.11	6.10
Procedures – termination by the employer In writing by hand, actual, special or recorded delivery	8.2.3 1.7.4	8.2.3 1.7.4	6.2.3
Architect (employer in DB2016) gives notice of "Specified Defaults" (a warning analogous to a yellow card in football)	8.4.1	8.4.1	6.4.1
If contractor continues with specified default for 14 days, employer may terminate employment within 21 days (a red card using the football analogy) For MWD2016, periods are shorter – 7 days continuing with specified default and 10 days termination thereafter	8.4.2	8.4.2	6.4.2
If the contractor repeats the specified default at any further time, the employer can terminate without a further formal time period	8.4.3	8.4.3	–

These warning periods need not be given for • Insolvency of the contractor – for the reason that quick and decisive action will be necessary to financially safeguard the employer • Corruption	8.5.1 8.6	8.5.1 8.6	6.5.1 6.6
Consequences of termination The employer may get others to complete the works, using available materials and equipment. The extra cost of completion, direct loss and expense is charged to the contractor	8.7	8.7	6.7
The employer may decide not to complete the works, in which case the extra cost of not completing as a result of the termination is charged to the contractor	8.8	8.8	–
Termination by the contractor, or either party for external events Similar to where the employer terminates except that the works are not completed. The contractor can claim payment up to date and any direct loss or damage (but only the latter, where the employer has caused the termination)	8.12	8.12	6.11
Termination where insurance cover for terrorism is withdrawn by insurers The possibility of the parties wishing to terminate the employment of the contractor in the event that buildings insurers withdraw cover for terrorism is dealt with separately in SBC/Q2016 and DB2016. In this event, either party may consider it too imprudent to continue with the project The same provisions apply as with termination by either party (i.e. cost paid to contractor excluding direct loss and/or damage). Provisions are contained in the "Terrorism Cover" part of the contract (Insurance Section 6, not the Termination Section 8)	6.11.2.2	6.11.2.2	–

Destruction of the employer's existing structure			
The possibility of the parties wishing to terminate the employment of the contractor in the event that the employer's existing structure is damaged (e.g. by a specified peril – fire flood etc.) is also dealt with separately in SBC/Q2016 and DB2016. Damage during the alteration, extension or refurbishment of an existing structure might make it difficult or pointless to rebuild as existing (the opportunity might arise for a completely different building design). In this instance, the same provisions apply as with termination by either party (i.e. cost paid to contractor excluding direct loss and/or damage). Provisions are contained in the Insurance part of the contract (Insurance Section 6, not the Termination Section 8)	6.14	6.14	–

Safeguards

Because of the serious effect of a termination, there are safeguards built into the contract covering the need to be sure that the act complained of is truly fundamental (i.e. if it is a continuing default, it usually has to be repeated!), and to be sure that the intention to terminate is communicated to the other party (i.e. by insisting on the use of controlled, verifiable communication).

The financial consequences of termination

The financial consequences of termination are that the aggrieved party is recompensed on the basis of the contract rates wherever possible. Thus, the general law remedies – setting aside of the contract and the use of quantum meruit – are avoided. The aggrieved party can also claim, through the contract, the direct costs of the termination. These would include delay costs and loss of profit, limited only by the general legal rules of remoteness of damages.

Where a contractor is the defaulting party, the calculation of the costs of termination poses some problems. It involves comparing the actual total cost of the project (using another contractor as appropriate) with the costs had the contractor carried the job through to completion. Thus, two final accounts need to be prepared – one for the total works including the completion works and one for the original contract valued as if it had continued to completion (i.e. with variations etc. priced at the original rates). For an example of the calculations involved, see Appendix 1.

Where the contractor terminates, or where either terminate for an external event, there will be no point in carrying on with the project. The financial losses include the costs of recovering plant, machinery and temporary works. Any exceptional costs, such as standing time for plant and loss of profit, may also be claimed. Loss of profit will depend on whether or not the contractor can redeploy resources to other equally profitable projects. Loss of profit (referred to as "direct loss and/or damage" in the contracts) is not allowed, where termination is for an external event, unless the event was rooted in the employer's negligence. This is to avoid the situation arising of a contractor being rewarded with profit for an event it, or a third party, may have caused.

INSOLVENCY AND BANKRUPTCY

Bankruptcy happens to individuals and partnerships and is relevant to sole traders and unincorporated firms. The individual or partnership can be sued for a debt to the extent of his/her wealth, jointly or severally. If the debt cannot be paid, the individual or individual partners will be declared bankrupt and a Trustee will be appointed by the court to oversee the financial affairs of the bankrupt. The non-payment of a debt (minimum of £5,000) gives creditors the right to apply for bankruptcy invoking the provisions of the governing legislation – the Insolvency Act 1986[2].

All the bankrupt's assets (house, car, goods, money etc.) will be distributed for the benefit of the creditors of the bankrupt. The bankrupt will be allowed to retain necessaries such as clothing, limited personal effects, tools of trade and a vehicle if needed for business/work. The bankrupt will be disqualified from getting credit and holding assets until debts are paid, or for a period (usually one year). After this period, the debts are waived. Nevertheless, a bankrupt's credit rating will be severely affected for a number of years after the bankruptcy.

In bankruptcy, there is no real distinction between the individual and the bankrupt firm. Individuals are made bankrupt, but in the case of a firm holding itself out as a partnership, all the individual partners are equally liable for the debts of the firm. The debt is not discharged on the bankruptcy of one of the partners, irrespective of the terms of a partnership agreement. Creditors can look to all the partners for payment and, if necessary, all partners will be bankrupted in discharging the debt. This is what is meant by the term joint and several liability. The Limited Liability Partnerships Act 2000[3] allows for limited partnerships, where the liability of partners can be capped at the contribution of each partner to the assets of the firm (i.e. not joint liability and not beyond the value of the assets of the firm). Limited Liability Partnerships have a greater duty of disclosure than unlimited partnerships (have to lodge accounts and information with Companies House).

Liquidation generally refers to incorporated firms and is dealt with under a different part of the Insolvency Act 1986. A liquidator is appointed to deal with the assets of a company and similar provisions apply covering liquidation for the non-discharge of a debt. For non-payment of a debt of £750 or more, the creditor presents a **winding up petition** to the court and, if successful, the court will make a **winding up order**.

In liquidation, a distinction is made between the firm and the employees. The firm will have all its assets distributed to the creditors in accordance with clearly defined priorities, but the employees (subject to them not being involved in wrongful trading) will be immune from personal liability. Thus, the directors of a limited liability company, as employees, are protected. Directors holding shares may lose the shares!

Regarding the priority for paying the undischarged debts of a company from remaining assets, the important point to note is that all normal trade creditors are near the bottom of the list for any payments. Only the shareholders of the company rank lower in distribution of assets. In the case of a contractor becoming insolvent, the client, materials suppliers and subcontractors are all trade creditors. They are all treated equally in the liquidation and none can claim preference. Should attempts be made by any one party to gain preference (say by an employer paying a subcontractor directly instead of paying the amount to the contractor), the liquidator will require that the amount be paid again to him/her. Neither can a client unilaterally get rid of a liquidated or bankrupt contractor on the act of insolvency. The liquidator has the power under the Insolvency Act to continue a contract for as long as is in the interests of the creditors as a whole. As statute takes precedence over contract terms, these powers cannot be overridden by the contract in use and the clauses on termination covered above may not automatically apply.

Liquidator. This is a general term for the individual responsible for administering the bankrupt's estate, or the assets of a limited company on behalf of all creditors and within the law. For bankruptcies, the individual is also referred to as the "trustee in bankruptcy". For all liquidations, a government official, the "Official Receiver", has overall responsibility for investigation and for most will be the only appointment. For some bankruptcies and liquidations, however, a private insolvency practitioner may also be appointed. This will normally be for larger bankruptcies and liquidations, particularly where there is a possibility of rescue on restructuring.

"Administration" also applies to incorporated companies. A court can appoint an insolvency practitioner as an "Administrator" to take control of the company and to rescue it as a going concern or sell the business or assets. Creditors' rights to enforce payment of their debts are suspended whilst the Administrator is in office (a moratorium). A creditor, the shareholders or the directors can apply for the appointment of an administrator. An Administrator has wide powers and may wish to continue with existing contracts.

Duties of the consultants on liquidation of a contractor

Some actions after liquidation must be prompt – losses to an employer will continue and may not be recovered from the liquidator. However, other actions, such as arranging a continuation contract with another contractor, although urgent, are not so critical. Finally, sorting out the financial consequences of termination cannot take place until costs are known for both the failed contract and its replacement.

Immediate action

INFORM INTERESTED PARTIES

This will include the employer, sub-contractors, any **bond** provider and the contractor itself (so that it is aware of the termination action to be taken by the employer).

STOP PAYMENT

Even where authorised by certificate or statement, it is possible to suspend payment immediately on the act of liquidation. The cost to the employer of termination will probably outweigh by a considerable margin the outstanding payment and, in a process of "mutual dealing", it is legal to offset due payments against debts. Consultants will action suspended payments by advising the employer.

SECURE THE WORKS

As many operatives and subcontractors may not have been paid, the site materials and in some cases the finished work will be at risk of removal. In liquidation there is a great temptation to try to recover goods before ownership has been properly established. The works must, therefore, be properly secured and policed.

ARRANGE INSURANCE OF THE WORKS

Insurance for the works, site materials and any plant to be retained will be needed. This insurance may have lapsed on the liquidation of the contractor. As it will be necessary for the employer to take out insurance, the basis of insurance will have changed. Insurance companies may treat the part completed works as an existing building, with insurance of the work by a completing contractor being "works insurance". It will probably be more efficient for the employer to insure both the part completed building and new works (e.g. under Schedule 3 option C of JCT SBC/Q2016) and not to require an incoming contractor to insure (see also the section on insurance).

CONFIRM TERMINATION

In the event of the bankruptcy or liquidation of a contractor, it is usually better to have a clean break and start with a new contractor. By statute, a liquidator can elect to continue the contract, but this is not normally done as he/she will not have the necessary expertise to handle the running of a building contract. The liquidator would also have great difficulty in keeping the construction team together for a temporary period. However, it is necessary for the client to obtain confirmation from the liquidator that the contract will not be continued. It is also necessary for the client to go through the formalities of the contract itself and terminate the employment of the contractor under the appropriate clause (SBC/ Q2016 clause 8.5). Should there be a realistic prospect of a **novation agreement** (see below), it may be better to suspend termination temporarily whilst this possibility is explored.

CARRY OUT AN ACCURATE SITE SURVEY

This will usually be done by the Quantity Surveyor for SBC/Q2016, an employer's agent for DB2016, or the Architect for MWD2016. It will involve an accurate valuation of work carried out to date (most important as this will be the starting point for any completion contract), a detailed inventory of site overhead items including cabins, cranes, scaffolding etc., and an inventory of materials on-site. No site overhead items or materials should be allowed off-site until ownership is verified (see below).

Medium-term action

ARRANGE A COMPLETION CONTRACT

A choice on how the remaining works are to be carried out will need to be made. This is most likely to involve the introduction of a new contractor or contractors but could involve negotiating a **novated contract** with an insolvency practitioner.

NOVATION

A novated contract continues the employment of the original contractor, but as a new **novation agreement**. For some acts of insolvency, there may be prospects of rescue and it may be in the employer's interest to keep the contract going (e.g. with an arrangement for paying debts approved under the Insolvency Act 1986).

The contractor will be represented by an insolvency practitioner administering the company, who will be personally bound by the terms of the novation. The terms of the novation agreement are subject to agreement and could be:

1. Continuation of the contract, with the intervention of the insolvency practitioner. This would be appropriate where the insolvency event was a temporary set-back for the contractor but is unusual.
2. Pure novation involving agreeing to continue the contract on exactly the same terms as the old agreement. This might be possible early in a financially sound (from the point of view of the insolvency practitioner) contract, where a refloated firm will continue working.
3. Conditional novation involving varying the terms of the old agreement, perhaps by increasing the duration of the contract or amending some of the contract rates. Again, a refloated firm will be formed to take over the original contractor's duties.

In all three instances, the insolvency practitioner will be reluctant to accept personal liability for performance of the contract. As a refloated contractor is involved for (2) and (3) above, only (1) is really a novation and will only occur where the insolvency practitioner can help with re-establishing the contractor and then quickly step aside.

COMPLETION CONTRACT OPTIONS

This is rather like making a choice of procurement path for a new project. In the case of a contract nearing completion, it may well be best to let the work on a cost reimbursement basis. This will allow work to start quickly and avoid difficult problems of valuing work (i.e. the split between the new and old contract). Once structural work and cladding is complete, construction management may also be appropriate, provided that the client and advisors can handle the multiplicity of contracts involved.

For contracts where a substantial amount of work remains to be done, it will usually be better to arrange a lump-sum completion contract. This is done (in the case of a contract with Bills of Quantities) by producing a "completion bill" consisting of the original bill with the quantities adjusted to reflect the amount of work carried out by the liquidated contractor. An accurate valuation has to be made by the Quantity Surveyor to include an assessment of work completed, defective works and site materials. The completion bill is priced in competition or by negotiation and a "completion contract" is let. For contracts originally let using design and build or similar contracts, there will be little point in engaging another design/build contractor and it is more efficient to change procurement to a traditional or construction management form, using the original designs and, if available, the original architects as consultants.

Re-letting the contract takes a considerable amount of time and delay often forms a major expense. To reduce this it may be prudent to start some works of tidying up and protection on the basis of a cost reimbursement contract prior to letting the full completion contract. Occasionally, time can be saved where the liquidated contractor is acquired by another firm and the team is kept intact but under another name. In this case, the client will effectively be treating the new firm as a completely separate entity from the old and will have to agree new contract terms and amount. The problem with all completion contracts let on a lump-sum basis is that of the interface between the old and new work. The completion contractor has a readymade excuse to blame all defects in the works on poor preceding works and to try and claim as much of the value of all work as its own. Hence, there is a need for detailed assessment and valuation by the consultant team. Additionally, completion contracts are not popular with contractors as they involve poor relations with retained subcontractors (who may not have been paid for earlier work) and problems of picking up other people's work. This unpopularity leads to higher tenders for completion work.

Final account action

FINANCIAL ADJUSTMENTS

The production of the final account is very much complicated by liquidation. As far as the cost to the client is concerned, the gross cost will amount to the sum of the certificates issued and paid to the original contractor, the final amount due to the completion contractor(s), the cost of providing temporary works and protection, the cost of ensuing delay and the cost of any other incidentals such as professional fees, insurance and additional administration. The client will look to the liquidator of the original contractor for the difference between the gross cost and the cost of the original contract had it run to completion. As the original contract never did run to completion, the account for the work is a fiction (technically "notional"). The "notional account" is calculated by reworking the completion account with prices taken from the original account and the difference between actual and notional accounts has to be agreed between the employer and the liquidator (see Appendix 1 for an example of the calculations involved).

On agreement, the client will have proved in the liquidation and stands to receive a fraction of the debt outstanding pro-rata to the total of all the debts of the contractor. The amount paid also depends on the value of any assets remaining in the hands of the liquidator after all other prior charges have been met.

DEALING WITH PERFORMANCE BONDS[4]

A bond is effectively an insurance policy for a fixed amount (usually about 10% of the value of the contract), payable in the event of non-completion of the contract by the contractor. It is required by some clients (in particular public authorities). The bond issuer underwrites the cost of completing the contract, using other contractors if necessary, up to the limit of the bond. The amount of the difference between actual and notional costs will need to be negotiated between the client and the bond issuer for a claim on the bond to succeed.

Difficult areas related to liquidation

Ownership of materials subsequent to liquidation

Ownership of goods normally passes on delivery and thereafter a supplier can sue for payment but cannot retrieve the goods. Therefore, a contractor might own materials on-site without having paid for them as yet. This becomes a problem on liquidation as the supplier will be much more interested in retrieving the unpaid goods that receiving a fractional payment (often less than 1% of the value) from a liquidator. It is for this reason that materials suppliers usually include express "reservation of title" clauses in their supply contracts. These state that ownership will not pass on delivery, but only on payment. A reservation of title clause might mean that a liquidator could not seize materials on-site for the benefit of all the firm's creditors (including the employer). Neither could an employer seize the goods nor hold them as an asset pending settlement of accounts by mutual dealing.

More importantly, it is possible that materials on-site will have been paid for and will be expressed (e.g. in SBC/Q2016) as being owned by the employer. The materials, thus, have two potential owners – supplier and employer. Fortunately, once goods have been passed in good faith to third parties, reservation of title clauses cannot be applied and the supplier cannot retrieve the materials. This does not, however, stop suppliers from attempting to remove goods and adequate site security is important. As it is custom and practice for materials on-site to be paid for and hence owned by the employer, a liquidator cannot seize materials on-site, which have been included in an interim payment.

This is not the case for materials not on-site – perhaps still in the contractor's yard or stores. From the point of view of the employer, these are at threat from two sources – the supplier, who will possibly have a reservation of title clause in the supply contract, and the liquidator, who will seize the goods for the benefit of all creditors. To avoid losing materials to either of these parties, JCT/Q2016 and DB2016 have special requirements related to paying for materials off-site, including a requirement that they are owned by the contractor (i.e. without title being reserved to a supplier) and that they are clearly labelled with the employer's identity and the site (thus making it clear to a potential liquidator that they are not part of the general assets of the contractor).

"That which is attached to the land, goes with the land"[5], so once materials are built into the works they become the landowner's – in building contracts, the employer. It is, therefore, clear that materials fixed to the works cannot be removed by the contractor, or more likely subcontractors. However, the terms of some subcontracts are for labour and materials, with no payments on account, or for payment only on the basis of installed materials. In these contracts, ownership of materials will stay with the subcontractor until installed, when they become the employers. If an employer pays for such materials before being fixed, it will not gain ownership and, in the event of contractor liquidation, the subcontractor can retrieve them. To avoid this problem, an employer's agent should either insist that the contractor has paid for the materials in question before payment to the contractor (i.e. see receipted invoices) or (as required in SBC/Q2016 and DB2016) insist that all subcontracts make provision for passing of ownership of materials on delivery to site.

Re-engaging subcontractors

Liquidation of a contractor will also mean that all subcontracts will terminate. If the subcontracts are in similar terms to the main JCT contract (SBC/Q2016, DB2016), similar provisions to that outlined above will apply. Existing subcontractors will be under no obligation to continue working for either a new contractor or the employer directly.

This is clearly not a major problem, where the subcontract work has been completed (whether or not payment has reached the subcontractor), or where subcontract work has yet to start. In the latter case, a new subcontract will need to be agreed, but the subcontractor will not be owed payment from the liquidated contractor.

Where subcontractors are part way through their contract they will often be reluctant to continue working unless they are fully paid. The lack of an obligation to continue can be very problematic where:

1. The employer has already paid the liquidated contractor for work on account by the subcontractor, but the amount has not been passed on. The subcontractor will be treated as an unsecured creditor and will be left to prove for a fractional payment in the liquidation.
2. The subcontractor is installing a specialist or patent element, where work cannot be carried out by an alternative subcontractor.

In these circumstances, particularly where an employer is still holding amounts due to a subcontractor, it is tempting for an employer to pay the subcontractor directly and bypass the contractor. This preferential treatment will not be acceptable to a liquidator. It is the function of the liquidator to receive all debts owing to the contractor (including unpaid amounts held by the employer for subcontractors) and to distribute them to all creditors in proportion to the amount of indebtedness of each. If an employer does pay a subcontractor direct, it is within the power of the liquidator to demand payment from the employer to the contractor (in other words the employer will have to pay for the subcontract work twice).

This restriction also applies to retention held by the employer for all work (most of which will be executed by subcontractors). The retention is a debt owed to the contractor and should be paid to the liquidator. To avoid retention being claimed by a liquidator, JCT/Q2016 and DB2016 state that it is held in trust for the contractor and if requested is placed in a separate bank account. The employer cannot pay the subcontractor directly for unpaid retention, but the subcontractor may be able to claim it from a trust account without having to prove in the liquidation.

APPENDIX 1 – EXAMPLE TERMINATION ACCOUNT

Termination by employer				
Liquidation account				
			£	£
Original contract sum				1,200,000.00
Net adjustments (variations etc.) before termination				18,000.00
Subtotal				1,218,000.00
Tender for completion work			450,000.00	
Net adjustments (variations etc.) on completion work[6]			9,000.00	8,400.00
Completion final account			459,000.00	
Notional final account				1,226,400.00
Costs of termination				
Temporary and emergency work			2,400.00	

Additional professional fees		4,800.00	
Direct loss/expense (delay etc.)		9,600.00	
Total cost of completing		**475,800.00**	**475,800.00**
Total amount outstanding to original contractor			**750,600.00**
Less payments on account to termination			756,000.00
Amount due from contractor			**5,400.00**

NOTES

1. The Public Contracts Regulations 73.1(a)(c) at http://www.legislation.gov.uk/uksi/2015/102/introduction.
2. Insolvency Act 1986, 1986 Chapter 45.
3. Limited Liabilities Partnership Act 2000, 2000 Chapter 12.
4. For more detail on bonds, see Chapter 10.
5. *Holland v Hodgson* (1872) LR 7 CP 328.
6. Variations are valued at completion contract rates for the completion final account (paid to the contractor completing the work) and at original contract rates for the notional final account (would be paid to the original contractor had it finished the contract).

8

The supply chain and subcontracting

THE SUPPLY CHAIN – A CHAIN OF LIABILITY

Constructing a building is a complex assembly process where the eventual user is often remote from the commissioning owner and contractor. The contractor is usually supplying managerial expertise and employing specialists to carry out elements of work. In turn, these specialists may sublet work further. The materials and components that make up the fabric of the building are supplied by manufacturers directly or through builder's merchants. Some materials are traditionally supplied directly to the contractor, whilst others will be supplied to specialists.

In all transactions, from user to manufacturer, a contract of some form or other will be used.

- Between the user and current owner of the building, this transaction may be a tenancy agreement, which is a transaction of land.
- Between current owner and commissioning owner, there will have been a sale agreement – another transaction of land.
- Between commissioning owner and contractor, a contract for building will have been used (perhaps based on the JCT standard form SBC/Q2016).
- Between contractor and specialist, a contract for building a specialist element will have been used (perhaps based on a standard form derived from SBC/Q2016).
- Between specialist and materials supplier, a contract for supply (normally based on supplier's standard terms) will have been used.

The problem of breaks in the chain of liability

The nature of this chain of contracts becomes important when things go wrong, in particular in relation to latent defects. Latent defects are defects in the finished building that either have not occurred or had not been apparent at the time of construction. It may be that they do not appear as a problem for many years, but their cause relates to some failure in construction or materials manufacture.

Liability for the failure would properly lie with contractor, specialist or manufacturer and the user affected by it will be looking for redress from one or other of these parties.

However, the user has no contract with the organisations responsible, only with the next party along the chain (in the case of the user, with the current owner). It is necessary to take action under this contract and leave it to the current owner to pass liability on to the original owner, then to the contractor, specialist and so on. Passing liability down the chain in this way presents particular problems with buildings and construction. Because of the nature of the various contracts used in the chain, breaks occur and often act to prevent liability reaching the perpetrator of a defect. Using as an example a commercial building such as a small warehouse or industrial building and assuming the contracts in the simple chain outlined above, the major issues are as follows.

User to current owner – the FRI lease

Owners of commercial buildings are often major investment companies, holding the building to provide a long-term financial return. They will seek to minimise involvement in any work to a building they own and do this by organising the transaction of land (commercial tenancy) so that the tenant (user) is fully responsible for insuring and repairing the building (the FRI lease). Therefore, if any defect occurs in the building, the tenant will be responsible for its repair and cannot ask the owner to put it right. In addition to this, the tenant will (in theory) have had the opportunity to inspect the finished building and to uncover any defects prior to signing the tenancy agreement.

Current owner to commissioning owner – sale of land

Buying a completed building, as opposed to commissioning a building from a contractor, is a transaction for the sale of land and all that is permanently fixed to it. In land sale transactions "let the buyer beware" (caveat emptor). Buildings are sold with all faults and it is the responsibility of the buyer to carry out detailed inspections and uncover all defects. Once the building is sold, the buyer cannot normally go back to the seller and ask that defects be corrected or recompense be paid.

Commissioning owner to contractor – a contract to build

Larger building contracts are let using detailed contract terms allowing for redress in case of defects appearing after work is completed. However, redress is not unlimited. A defect must relate to a failure in building or manufacture of a component. Normal wear, tear, maintenance and replacement would be excluded. Nor would the period of time to take action be unlimited. Action is time barred after 6 years from the date of breach of simple contract. This date is usually the date of agreed final completion – in SBC/Q2016 the issue of the final certificate by the architect. For deed contracts, action is time barred after 12 years from the date of breach.

The actions of the commissioning owner (Employer, in the language of the JCT contracts) and agents such as an architect can also remove liability from the contractor. Examples illustrate this:

1. The Employer directly commissions work to the building alongside the main contract work. The contractor would not be liable for defects in this work.
2. An architect requires that materials from a particular manufacturer are used in constructing the building. The contractor may not be liable for defects in these materials.
3. An architect requires that certain work be carried out by a particular specialist firm. The contractor may not be liable for the quality of that work.
4. An architect expresses satisfaction with work carried out to his or her approval. The contractor may be able to claim that this expression is proof that work was satisfactory.

In all these instances, the Employer may not be able to pass liability along the chain to the contractor. However, for 2, 3 and 4, there may be an alternative of taking action against the architect if he or she carried out duties negligently.

Contractor to specialist – a contract to build specialist elements

As contractors are normally providing managerial expertise and limited physical work, it is important that terms of the subcontracts they use with specialists are back to back (closely match) with those of the contract between employer and contractor. For all the major JCT forms of employer-contractor main contract, there are corresponding JCT subcontracts, where wording is often identical to the main contract. It is, therefore, fairly easy to pass liability from contractor to specialist. There are a limited number of issues that might arise only between contractor and specialist, but these are not relevant to the main contract and will not affect those further up the chain.

Specialist to supplier

Building component manufacturers and builder's merchants are often large and powerful commercial firms supplying to the whole industry nationally and internationally. They can, therefore, impose their own standard terms of trade on their buyers. These terms are weighted towards their own interests, in particular containing reservation of title and limitation of liability clauses.

The latter is of particular relevance to obtaining adequate redress for latent defects. Suppliers typically limit liability to replacing the component concerned. However, the cost of replacement can be far higher than this once the component is installed in the building and may include the cost of:

- Extracting the component from the work
- Taking down sequential work to obtain access to the component
- Installing a replacement component
- Delay to the work in question, sequential work and, possibly, the overall project
- For replacements once the building is occupied, disruption to activities of the user

Solutions

The upshot of the various breaks in the chain of liability is that it can be very difficult from the outset for a user to obtain redress when the building fails to perform as expected. To deal with this lack of effective redress, a number of solutions have been developed:

- Collateral warranties
- Use of the Third Party Rights Act 1999[1]
- Taking action using the law of tort
- Avoiding certain actions removing liability from contractors

Collateral warranties

A collateral warranty is a contract running alongside another (hence the name "collateral"). It is possible to set up a direct contract, for example between the tenant for the above building, and any of the building team. This contract is designed to deal with breaks in the chain of liability in relation to common problems, especially latent defects. A typical term might be that a specialist warrants that it has designed and installed a specialist element with due skill and care. If the element fails, the tenant will be able to take action using the warranty. In English Law, it is necessary for some benefit to be conferred on each party by the contract and this benefit is known as "consideration", but in this case, the specialist receives no consideration additional to that received under the separate supply chain contract with the contractor. For collateral warranties, therefore, it is essential that some nominal payment be made to the warranting party. This nominal payment might be a sum of money (£1.00 for the JCT standard warranties mentioned below) or might just be the traditional "peppercorn" on the legal principle that consideration must be real but need not be sufficient.

With many JCT standard forms of contract, including SBC/Q2016 and DB2016, there are now matching standard forms of warranty related to purchasers, funders and tenants of buildings (the benefiting parties) and both contractors and subcontractors (the providing parties). The funder would often be the current owner, who at the time of construction had not actually purchased the building. There is also a facility for a standard warranty between employer and subcontractors. The full list of standard warranties as envisaged by SBC/Q2016 is:

1. CWa/P&T – A direct warranty between contractor and purchaser or tenant
2. CWa/F – A direct warranty between contractor and funder
3. SCWa/P&T – A direct warranty between subcontractor and purchaser or tenant
4. SCWa/F – A direct warranty between subcontractor and funder
5. SCWa/E – A direct warranty between subcontractor and Employer

Clauses 7C-E of SBC/Q2016 and DB2016 enable the use of the appropriate warranty, which should be included in the Contract Particulars and in the tender bills of quantities (or Employer's Requirements for a Contractor's Designed Portion – CDP). In addition to warranties mentioned in the JCT contracts, it is also common for collateral warranties to be used between benefiting parties and consultants – architects, engineers, services engineers and quantity surveyors. Note that there are no express provisions in MWD2016 covering collateral warranties and should they be required, separate provisions would need to be incorporated in tender and contract documents.

Given the large number of subcontractors used on even a fairly simple project, it can be seen from the above that there could be a correspondingly large number of warranties. In practice, however, the use of warranties would be limited to specialist elements where latent defects are most likely – cladding and roofing, services installations and other elements where the specialist was involved in both design and construction. For architect or engineer designed work, particularly where building control inspections are involved, there is less need for direct contracts with specialists as failures should be less likely and warranties can be taken out with the consultants.

Collateral warranties extend the liability of the providing parties and thus increase costs. It is, therefore, important that all providing parties are given notice in tenders and consultancy appointments that warranties will be required. If not, parties might refuse to provide the warranties or ask for additional payment.

Use of the Third Party Rights Act 1999

This Act is designed to make it easier for a third party (e.g. purchaser, tenant or funder) to take action directly against a liable party under the principal contract, without needing to use a collateral warranty. Colloquially, the mechanism involved is known as "shoe stepping" – the third party is enabled (in the principal contract terms) to step into the shoes of the principal contract beneficiary and take action. For example, the user mentioned above could be enabled in the contract between employer and contractor to use the terms of the main contract to sue the contractor.

Such terms have been used in the past but have encountered the problem that the liable party has been able to contend that the principal beneficiary (employer) has lost nothing as the result of the breach of contract – by the time the latent defect has manifested, the employer has sold the building! The Act prevents this contention.

For the Act to work, the benefiting third party must be mentioned by name or class in the principal contract. SBC/Q2016 and DB2016 deal with third party rights in Clauses 7A and 7B. 7A allows third party rights for purchasers and tenants. 7B allows for third party rights for a funder. Similar provisions can be used in subcontracts and sub-subcontracts further along the chain of liability, although it is unlikely to be incorporated in the standard supply conditions of materials suppliers and manufacturers. There are no corresponding express provisions in MWD2016 and, if required, they would need to be written in as amendments.

Taking action using the law of tort

A tort is a civil wrong and the law of tort is a branch of law separate from contract. Any person has a general duty to take care not to injure someone who is in close proximity (a fellow user of the road, in a motoring accident, or a neighbour in a domestic situation). In the past, breach of a duty to take care has been used to take action in commercial cases where economic loss was involved and this remains a possibility. For example, a tenant discovering a leak in a roof, which damages the contents of a building, might sue the roofing specialist in tort.

Consideration of the law of tort and its application to building work is beyond the scope of this book, but in general, such action is uncertain of outcome and expensive. The courts in the past have declined to allow this route to redress where purely economic loss was involved, but difficulties surround what exactly is "economic loss" – all losses ultimately being reducible to monetary values. The practical advice in relation to contemplating this course is to attempt to avoid it in the first place by using collateral warranties or third party rights and keep it as a last resort when fighting for a principle.

Avoiding certain actions removing liability from contractors

As a general principle it is unsatisfactory to split liabilities for the adequacy of an artefact (or even whole building) between separate parties. It is not, therefore, recommended for an employer to directly take over elements of constructing a building and have them carried out separately within the same time period as the building contract. This is unlikely to be problematic for small, discrete elements (e.g. a feature sculpture at the entrance to a building), but for more major works, such as fitting out an office, it is probably better to wait until the building is finished. This will avoid the employer having to take responsibility for the delay and disruption caused to a contractor by the direct works.

Similarly, as a general rule it is better for an architect not to specify materials in such a way that the contractor has no option but to use a specific manufacturer. If this is done, the employer would not be relying on the skill of the contractor in selecting the specific material and the contractor would not be held liable if the material failed.

The same consideration applies where an architect or other agent (e.g. Employer's Agent in DB2016) requires that certain work is carried out by a particular specialist firm. The employer would not be relying on the skill of the contractor in selecting the firm and the contractor would not be held liable if the firm failed. If the firm is involved in design, the employer would have no redress for defective design through the principal contract. The only possible course of action would be to sue the architect for negligence in requiring the specialist appointment. Current thinking is that architects and employer's agents can select firms to carry out specialist work. They should give the contractor a choice of firms from which it uses its skill in selecting one – a minimum choice of firms should be three, but should preferably be more. In these circumstances, the contractor would be fully liable for the performance of the specialist.

Finally, architects should not demand that work be to their satisfaction, but rather should specify some objective measurable standard. The reason for this is contained in the wording of many JCT contracts, which state that the final certificate is conclusive that where work is to be to the architect's satisfaction, that satisfaction has been obtained. The effect of the language of the JCT contracts is to transfer liability for such work from contractor to architect.

SUBCONTRACTING

Introduction

Subcontracting is widely used in construction industries throughout the world. In the UK construction industry, it is common for modern contractors to sublet all work and employ only key managerial workers and a limited number of operatives to deal with minor snagging and maintenance work. Subcontracting is normally permissible for construction and the contractor does not have to carry out all work itself. The identity of the people carrying out the work is not essential to performance. This should be contrasted with architecture, where the individual identity of the architect is often important to his or her client and the latter could expect at least some input from the chosen individual. Although work may be subcontracted, the contractor will remain fully liable for carrying it out.

In JCT contracts, subcontracting is controlled by requiring consent:

Contract	From whom	Clause
SBC/Q2016	Architect	3.7.1
DB2016	Employer	3.3.1
MWD2016	Architect	3.3.1

For subcontracting the design, in DB2016 and contractor designed portions in SBC/Q2016 and MWD2016, consent is separately required, presumably because the execution of design may be more particular to the contractor.

Even within subcontract organisations, it is common for work to be sub-subcontracted, often down to the level of the self-employed worker. This arrangement has a number of advantages:

- Increased productivity as the subcontractor is effectively on a 100% bonus system.
- Risk transfer from the contractor to the subcontractor – this allows tighter pricing and a more certain return.
- Lower employment overheads.
- Some tax advantages at the level of the self-employed worker.

Disadvantages include:

- Less direct control of quality – control has to be exercised through a commercial contract rather than a contract of employment using direct incentives.
- Poor apprenticeship arrangements – self-employed people are likely to be less keen on taking on trainees.
- Poorer welfare benefits for the operative.
- Less scope for a structured promotion system feeding into managerial posts.

Changes to the tax arrangements for the industry have made it more difficult for contract workers to claim self-employed status. If they work for only one company, they will be regarded as employed by that firm and be subject to tax deductions at source. This has reduced the number of self-employed labour-only individuals in the industry. However, between bona-fide firms, subcontracting will continue to feature.

Terminology

A variety of terms are used to describe subcontractors, depending on the form of contract and the level of the subcontract.

- *Specialist*: A firm specialising in a trade or element of construction. A general term for subcontractor, works contractor etc.
- *Subcontractor*: A specialist contracted to a contractor (often termed a "main contractor") who in turn is contracted to the employer.
- *Sub-subcontractor*: A specialist contracted to a subcontractor. Subcontracting sometimes extends further (sub-sub-subcontractor!).
- *Domestic subcontractor*: An archaic term for a subcontractor used to differentiate the firm from a **nominated** or **named subcontractor**. The term was used in earlier versions of SBC/Q2016. Included here for the sake of completeness.
- *Nominated subcontractor*: An archaic term for a subcontractor (typically used in earlier versions of SBC/Q2016) selected by the architect and for whom the main contractor *is not* fully liable to the Employer. Nomination in some form is likely to remain a feature of British-style contracting.
- *Named subcontractor*:
  1. A subcontractor (in SBC/Q2016 and earlier versions) selected by the contractor from a list of at least three named in the Contract Bills and for whom the contractor is fully liable to the employer.
  2. A subcontractor in DB2016 (the Design/Build Contract) selected by the employer before letting the main contract and for whom the main contractor is fully liable, with exceptions. The Employer names the subcontractor in the Employer's Requirements and Schedule 2 Part 1 Section 1 (Supplemental Provisions). The exceptions are:
     - Where the contractor is unable to appoint the named subcontractor, the employer may require the contractor to carry out the work, or arrange to employ a specialist directly but is liable for the cost and delay involved.
     - In circumstances where the subcontractor defaults (e.g. by becoming insolvent), the cost of the default is treated as a change and added to the contract sum.
- *Named specialist*: A subcontractor in SBC/Q2016 similar to a named subcontractor in DB2016, named in the contract documents, but may be named post-contract against a provisional sum.
- *Works contractor*: A subcontractor (in MC2016 – the management contract) selected by the architect and management contractor and for whom the management contractor *is not* fully liable to the Employer (see below).

- *Trade contractor*: A specialist (in CM/TC2016 and CM/A2016 – the construction management contracts) in connection with construction management. It is not really a subcontractor as the contract is directly between the employer and the trade contractor.
- *Labour only subcontractor*: A subcontractor providing labour but not materials. In the UK, this is usually only for the traditional building trades such as brickwork, carpentry and plastering. Materials are supplied by the contractor.
- *Labour and materials subcontractor*: A subcontractor providing labour and materials. In the UK, this is usually for less traditional building trades such as services, steelwork and patent cladding but may also include traditional trades where larger subcontractors are engaged.
- *Novated subcontractor*: A subcontractor (usually a design firm) where:
    1. The main contractor has agreed as a condition of the award of the main contract to appoint the design firm as a subcontractor. The main contract is usually design and build (DB2016).
    2. The design firm has agreed as a condition of the award of the pre-contract design contract (with the Employer) to subsequently act as a subcontractor to the main contractor.

Specialists in JCT contracts (standard JCT forms given next to contract lines)

SBC/Q2016

Sub-Contractors and Named Sub-Contractors

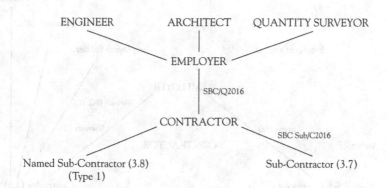

SBC/Q2016

DB2016

Sub-Contractors and Named Sub-Contractors

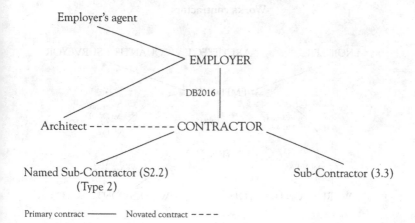

DB2016
Warranties only: CWa/P&T, CWa/F, SCWa/P&T, SCWa/F, SCWa/E

Standard JCT warranties (DB2016)

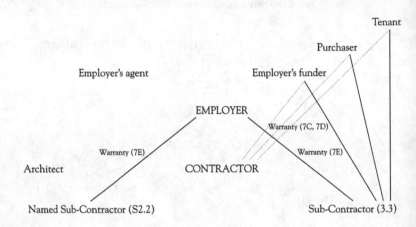

MC2016

Works contractors

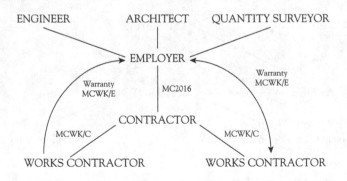

CM/A2016, CM/TC2016

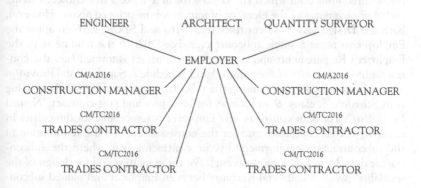

Problems with subcontracting

Back-to-back problems

The problem of a mismatch between the terms of contracts for main and subcontracts has been dealt with in the section on the supply chain. It is important, where an owner wishes to take action to get defective work rectified, that liability can be passed effectively along the chain to those ultimately responsible – that contracts are fully back to back with each other. Passing liability has been impeded by the practice of architects naming or nominating single specialist firms to carry out specific work and this practice is still expressly allowed in SBC/Q2016 and DB2016. Nominating single firms is still informally practiced, especially for smaller works.

By naming or nominating a single specialist, the employer is not relying on the skill of the contractor in selecting the specialist but is, rather, imposing the specialist on the contractor. Specifically, case law has indicated that nominating a subcontractor relieved a contractor for liability towards the employer for "4 Ds":

i. Delay by a nominated subcontractor
ii. Default of a nominated subcontractor
iii. Design by a nominated subcontractor
iv. Delay in designing by a nominated subcontractor

A similar problem arises where materials are specified by trade name or in such a way that an architect or employer reserves the right to make a selection. Neither architect nor employer is relying on the contractor to provide goods, which are fit for purpose; therefore, the contractor would not be liable to an employer if they fail. This creates a break in the chain of liability.

The traditional JCT standard form of contract SBC/Q2016 seeks to avoid nomination and, when naming firms, the architect is required to provide a list of three firms from which the contractor may make a free choice. Having such a choice removes the element of imposition mentioned above. However, both the Design and Build contract (DB2016) and SBC/Q2016 do allow the Employer to name a single subcontractor. For DB2016 the naming is in the Employer's Requirements and not once the contract is formed (i.e. the contractor has prior notice of the naming) and Schedule 2, Supplemental Provision 1 provides a procedure for naming the subcontractor. For SBC/Q2016 naming is in Schedule 8, clause 9 and allows for both pre- and post-contract "Named Specialists". For both contracts, the Employer is responsible for difficulties in engaging the subcontractor and for the consequences of valid termination of the subcontractor's employment by the contractor (e.g. where the subcontractor defaults or becomes insolvent). For these matters and for design of the specialist's work, a collateral warranty between Employer and named subcontractor, or specialist, is necessary.

There are no provisions for naming subcontractors in MWD2016. Further, the form is expressly stated as being unsuitable "where provisions are required to govern work carried out by named specialists". Nevertheless, agents such as architects often informally engage in naming or nominating single firms to carry out specialist works with this form and the above considerations apply. Where naming is essential, it is better to give the contractor a choice of at least three specialists, particularly where the specialist will be designing the work in question. If only a single firm will do, then a collateral warranty should be used to ensure direct liability from specialist to employer.

Set-off

Set-off is a counter-claim from employer to contractor, or, more usually, from contractor to subcontractor for late completion, delay, defective work etc. The term is used informally here, as legally, set-off has a more restricted meaning relating to reducing a claim in court action.

Typical heads of contractor – subcontractor argument include:

- Delay
- Attendance
- Defects and consequential re-working
- Programming and knock-on effects (usually on other subcontractors)

In the absence of express provisions, a common law right to set off exists, subject to three provisos:

   i. Must be specific and quantified
  ii. Costs must have been incurred – i.e. not prospective
 iii. Notice must be given

In practice subcontract terms are often tipped strongly in favour of the main contractor with the subcontractor being required to take responsibility for many of the above heads of argument, irrespective of fault. However, industry standard subcontract terms and the provisions of the Construction Act now define minimum requirements for setting off amounts (both at main and subcontract levels):

1. Costs to have been incurred
2. Costs to have been quantified
3. Notice to have been given to the subcontractor (in writing) of intended set off (A Pay Less Notice in the language of the Construction Act)
4. Time limits regarding the notice to have been complied with
5. Mechanism for dealing with disputed set off including:
    i. Mediation
   ii. Adjudication

Battle of the forms[2]

This is a particular problem with informal subcontracts and where would-be contracting parties form contracts by exchange of letters, orders, confirmation of orders, estimates and similar. SBC/Q2016 is formed on the basis of the "Contract Documents" as identified in Clause 1 of that form. This means, provided the contract is entered as intended, that all the requirements are set out at the start and there is little room for argument as to whether a contract has been formed and under what terms.

In contrast, subcontracts are often formed by exchange of letters or orders. A typical scenario is as follows:

- Invitation to tender – extract from bills of quantities sent to subcontractor
- Offer – estimate sent which refers to conditions, perhaps on the back of the offer document
- Counter offer – acceptance or "confirmation of order" sent with either changes to detail or changes to terms (in the form of referring to different conditions on the back – this time of the confirmation)
- Counter-counter offer – revised estimate or acceptance of confirmation sent again referring to conditions on the back of the document
- And so the process continues until work, an argument or both starts. If work starts before argument, it may be that there is an *unconditional* acceptance by the action of one of the parties (by starting work or allowing work to start).

If so, whoever "fired the last shot" in the battle of the forms wins and the terms are as "on the back" of his/her form. If the fight is still continuing at the time of the argument, the argument will often be over whether any contract is in existence

Liquidation[3]

Insolvency of a contractor can give problems for both employer and subcontractor. This applies in relation to:

1. Reservation of title of goods on site and paid for. Both the employer and supplier/subcontractor may claim ownership of goods.
2. Retained ownership where subcontractors have not been paid for materials under a supply and fix contract. For supply and fix contracts, where the subcontractor is paid on completion of the work, ownership of goods passes on *installation* of the goods, not on delivery to site. If the employer has paid for goods in question, both the subcontractor and employer may claim ownership.
3. Payment direct of subcontractors on insolvency of a contractor. The employer cannot give any creditor preferential treatment by paying it direct. The official receiver or an insolvency practitioner can insist that payments contractually channelled through the contractor are paid to him/her.
4. Payments to subcontractors of retention held by the employer. Subcontractors have difficulty in obtaining retention, even though the subcontract may state that it is held by the contractor in trust. Unless there is an identifiable sum of money set aside in a separate trust account for each subcontractor, the funds will be lost in the general assets of the liquidated contractor.

Mechanisms for subcontracting

The standard mechanisms developed for placing subcontracts have the object of reducing or removing the problems noted above, in particular.

Battle of the forms

Industry standard subcontracts are designed to clarify and standardise the terms under which the parties contract.

Set-off, unfair contract terms etc.

Again, industry standard forms are drafted by agreement with representatives of both contracting and subcontracting organisations and include fair terms. The Construction Act as amended regulates the most obvious abuses of power.

Back-to-back problems

Remedies to back-to-back problems include:

1. Using collateral warranties. To ensure that these warranties are not found ineffective for want of consideration, all are agreed on the basis of the appropriate beneficiary paying a nominal sum (currently £1). Clauses 7C-E enable the use of the appropriate warranty, which should be included in the Contract Particulars and in the tender BQ (or Employer's Requirements for DB2016 and the Contractor's Designed Portion (CDP) in SBC/Q2016).

   When naming a subcontractor in the Design and Build form of contract (DB2016), the use of a warranty between Employer and subcontractor is important to specifically cover default by the subcontractor. When naming a specialist in SBC/Q2016, the use of a similar warranty is needed to cover design, default and delay (caused by default) (see above).
2. Using the Third Party Rights Act 1999. SBC/Q2016 Clauses 7A allows third party rights for purchasers and tenants and 7B allows for third party rights for a funder. Schedule 5 (Third Party Rights) limits rights to a first, second or third purchaser, tenant or funder.
3. Relying on the law of tort to take action directly against a third party.
4. Not nominating, or naming a single subcontractor, but always providing a choice of three to the contractor if the identity of the subcontractor is important to an architect or employer. SBC/Q2016 allows for the naming of three in clause 3.8, but note that equivalent clauses are not contained in DB2016 or MWD2016.

Arguments concerning attendance, programming and delay

These will continue, but some standard documentation tries to limit them (e.g. Management Works Contract Tender and Agreement (MCWK/T) requires express agreement between the Management Contractor and Works Contractor for these matters before the contract is entered).

NOTES

1. Contracts (Rights of Third Parties) Act 1999, 1999 Chapter 31.
2. For detail of "The Battle of the Forms" see Butler Machine Tool Co Ltd v Ex-Cell-O Corporation Ltd (1977) EWCA Civ 9.
3. See also Chapter 7.

9

Indemnity and insurance

INDEMNITY

Who needs and indemnity and why?

Someone is passing a building site (maybe an ordinary member of the public, maybe a famous celebrity or politician) and a brick slips off the scaffolding and falls on her head! Who does she sue for the injury and ruined hat (assuming that she is alive to take court action)? The bricklayer, bricklaying firm, contractor, scaffolding firm, architect, quantity surveyor, employer and more could all have had a hand in the incident:

1. The bricklayer: clearly culpable, but perhaps having little in the way of assets to pay out in a settlement.
2. The bricklaying firm: maybe liable as the employer of the bricklayer, but also often insubstantial and may be forced into insolvency by a large settlement.
3. The contractor: responsible for the overall coordination of all specialist firms and for organising the safety of the site, the contractor is likely to be liable. However, as with the bricklaying firm, the contractor may be driven to insolvency by a large settlement.
4. The scaffolding firm: responsible for providing features that make it difficult for bricks to fall off the scaffolding. If these features were missing, the firm may be liable but again could be financially ruined in an action.
5. The architect: designs the building under construction and also has supervisory duties to see that it is built as designed. These duties could include ensuring that the design can be constructed without danger to the public and inspecting to see that the work is being carried out to the design.
6. The quantity surveyor: quantifies the building under construction but also drafts preliminaries including general clauses requiring the contractor to comply with health and safety legislation and good practice. Failure to properly draft such clauses and see that selected contractors have the ability to comply could generate a liability.

7. The Employer: gives control of the building site to the contractor, design of the project to the architect and overall has little direct role at all in the proceedings. Nevertheless, the Employer is the building owner and, in the final analysis, remains (strictly) liable for what goes on there.

The answer is that she will be advised to sue them all (join them all in an action) and leave it to the judge to decide if all or any are liable, in what proportion each is liable and how much will be paid by each! A common scenario is that (2), (3), (6) and (8) have a liability and an example of the extent of liability for each in a successful £1m action might be:

Liable party	Proportion of liability (%)	Amount of liability (£)
Bricklaying firm	40	400,000.00
Contractor	50	500,000.00
Architect	5	50,000.00
Employer	5	50,000.00
Total	100	1,000,000.00

The Employer, having no active part in the project, might want some protection from other members of the building team against having to pay the £50k – this protection is called an **indemnity**. Having an indemnity is no protection should a party fail to pay (perhaps because the pay-out drives it into liquidation). The parties are "jointly and severally liable", which means that if one cannot pay the claim will be spread amongst the other parties. Assuming that the contractor goes into liquidation as a result of the action, a revised distribution would be as follows:

Liable party	Proportion of liability (%)	Amount of liability (£)
Bricklaying firm	80	800,000.00
~~Contractor~~	~~50~~	~~500,000.00~~
Architect	10	100,000.00
Employer	10	100,000.00
Total	100	1,000,000.00

The liability of each remaining party has increased proportionately to their original liability after excluding the liquidated contractor. In the extreme, a party, such as architect, or employer, could end up taking the whole liability should all the others fail as a result of the action! Therefore, not only is it prudent for the Employer to ask for an indemnity, it is also important to demand appropriate insurance from relevant parties should *bricks be dropped* or other third party liabilities arise.

JCT contracts only deal with liabilities between the Employer and Contractor. An indemnity sets out to make the Contractor solely responsible for acts or omissions under its control. It does not remove the Employer's liability to third parties, but, as between the Employer and the Contractor, it provides a mechanism to redistribute the liability. This redistribution can be problematic as it may go against some court decisions. If a court decides that a liability (however small) is owed by an Employer, then such an indemnity is thwarting the court's intent. Therefore, if there is any ambiguity in the wording of the indemnity, or if the indemnity is patently unfair, it might not be supported in an action (the clause will be construed "contra proferentum" – against the party drafting it). Additionally, if the Contractor cannot support the indemnity and pay up, perhaps because the extent of damages has caused its liquidation, the Employer will still be liable for any recompense. Worse, if several parties are held jointly and severally liable for a loss, under the rules of "contribution"[1] the Employer and other parties might find themselves picking up the cost of damages otherwise due from the Contractor.

How the contracts operate

In SBC/Q2016, clause 6.1–6.2 requires the Contractor to, first, indemnify the Employer against claims resulting from personal injury, and, second, to indemnify the Employer against claims resulting from damage to property:

Topic	SBC/Q2016	DB2016	MWD2016
The personal indemnity clause This requires the Contractor to indemnify the Employer against any proceedings except those arising from the Employer's own negligence. Very clear words indeed would be required in the contract if the Contractor was to indemnify for the Employer's negligence also and such a requirement might be considered unfair under the Unfair Contract Terms Act 1977[2]	6.1	6.1	5.1

The damage to property indemnity clause This requires the Contractor to indemnify against any proceedings arising *from its own negligence*. This indemnity is, thus, not as stringent as that in 6.1 above. If damage occurs to third parties and it is not the result of the Contractor's negligence, then the Employer will not be able to rely on this clause (but see 6.5.1 below)	6.2	6.2	5.2

Indemnities of the above type are only effective between the parties in contract and do not preclude the aggrieved third party from suing the Contractor, Employer, Architect, Quantity Surveyor, or any individual worker involved. As a claim for personal injury could easily make a contractor insolvent and leave the Employer open to an action, the contract incorporates mandatory insurance provisions.

Contract insurance

INSURING THE INDEMNITY

Topic	SBC/Q2016	DB2016	MWD2016
Insurance against personal injury and property damage The Contractor is required to insure against the liability arising from the indemnity clause	6.4.1	6.4.1	5.3
Contractor's insurance of liability of Employer (no negligence insurance) The contract provides that if the Contract Particulars (Employer's Requirements in DB2016) state that this insurance may be required and the Architect so instructs, the Contractor is required to insure against the Employer's liability to third parties not covered by the damage to property indemnity in clause 6.2 (the indemnity is only required in the event of the Contractor's negligence)	6.5.1	6.5.1	–

A Provisional Sum may be provided in the bills of quantities, or Employer's Requirements. In the final account, the Provisional Sum will be omitted and the cost of the insurance policy will be added to the contract sum			
This insurance is recommended in urban areas, where damage to adjoining property and equipment is possible. Possibly the hat of the "victim" mentioned above, but more likely subsidence and similar damage. Its requirement stems from the limited indemnity provided by the contractor for property damage. However, the exclusions to the insurance requirements are very widely drawn (e.g. to exclude damage which can be seen to be inevitable given the nature of the work). Again, the Employer has power to insure itself on the default of the Contractor			

INSURING THE WORKS

In contrast to insuring the indemnity, which is required to protect the Employer from third party action, insuring the works would normally be the Contractor's sole responsibility. In the event, for example, of a fire, the Contractor would still be required to deliver up the building on time and to the required quality. However, two circumstances lead to JCT contracts having detailed requirements for insuring the works:

1. Business prudency makes it desirable for the contract to have detailed works insurance provisions. A major uninsured event could induce the insolvency of the Contractor, leaving the Employer to complete the works. The contract, therefore, contains detailed provisions requiring insurance, powers to inspect documents and to insure on the failure of the Contractor to do so.

2. It is sometimes better for the Employer to insure the works rather than the Contractor. This is the case in circumstances where the Employer can get better insurance terms – for example where it is a professional client carrying out several contracts, or where it owns a large estate and holds insurance

already for other buildings. It is also likely to be the case for work in, or extensions to existing buildings. The Employer will already hold buildings insurance for the existing building and adding works insurance may be cheaper than asking the Contractor to take it out. The Employer insuring both existing building and works also avoids splitting liability for insurance between two or more insurers. Split liability could lead to arguments over which insurance company is liable for what loss.

SBC/Q2016 and DB2016 have three insurance options to cater for these circumstances, invoked in Clauses 6.7–6.10 and Schedule 3 of the contract. Only one option is selected and noted in the Contract Particulars. MWD2016 has two options and an alternative "Do It Yourself" third option, written directly into the contract wording.

Topic	SBC/Q2016	DB2016	MWD2016
Option A – insurance by Contractor The first alternative, providing that the Contractor is to insure the works on an all-risks basis, is the normal provision for new works	Schedule 3, Option A	Schedule 3, Option A	5.4A
Option B – insurance by Employer The second alternative is for the unusual situation where the Employer is to insure new works on an all-risks basis. This clause is intended where the Employer can obtain insurance on better terms than the Contractor, or in the Local Authority version of the contract, where the Employer can act as its own insurer There is no similar option in MWD2016, presumably because this situation is unlikely with smaller works	Schedule 3, Option B	Schedule 3, Option B	–

Option C – insurance of works, existing buildings and contents by Employer The third alternative is for the Employer to insure where the works are being carried out within, or as an extension to an existing building. The insurance is for three items: 1 The works, insured by the Employer on an all risks basis 2 The existing structure, insured by the Employer for specified perils (because all-risks insurance is not available for existing buildings) 3 The contents insured by the Employer in the existing building/new works	Schedule 3, Option C	Schedule 3, Option C	5.4B
Insurance of the works and existing structures by other means For MWD2016, the third option is to allow the parties to use whatever alternative insurance arrangements they wish. This is intended to allow for the situation where it is not possible or economic to arrange insurance of both existing structures and new works with one insurer. This may happen, for example, when the Employer is a tenant and buildings insurance is provided by a Landlord owner. Although providing flexibility, the option is not "fail safe" in the way that other pre-structured options are and arranging adequate cover in detail needs expert professional advice. Giving this advice would fall to the Architect or Contract Administrator in this contract			5.4C

Enforcement of insurance requirements To make sure all insurance policies are taken out, there are powers for the non-insuring party to enforce the requirements. For SBC/Q2016 and DB2016, this involves two stages: • Powers of inspection. • Powers to take out insurance in default. The cost to be deducted from or added to the accounts as appropriate For MWD/2016 there is no express power to take out insurance in default, but that would not preclude either party from doing so and claiming the cost outside the express terms Where the Contractor is expected to insure and a deduction is made by the Employer, the deduction will not show on an interim certificate and requires notice in accordance with the detailed payment provisions of the contracts and the requirements of the Construction Act	6.12	6.12	5.5
Payment and reinstatement Where the Contractor is to insure in accordance with Option A (5.4A in MWD2016), the Contractor gets only such amounts as are paid out by the insurer and such amounts are to be paid through the Architect (or Employer for DB2016)	6.13	6.13	5.6
Where the Employer is to insure the cost of repair and remedial works are treated as if they were a variation and added to the Contract Sum	6.13	6.13	5.6

Basis of risks Works insurance is expressed to be "All Risks" – that is the cover is for all risks except those expressly excluded, mentioned either in the contract (SBC/Q2016, DB2016 clause 6.8), or in the insurance contract (MWD2016)	6.8	6.8	5.4A
For existing buildings, it is not possible to insure for all risks and the alternative of insuring "Specified Perils" is provided	Option C	Option C	5.4B
Termination as a result of damage The three contracts give either party the power to terminate the Contractor's employment under the contract in the circumstances where damage to the works might mean that it is not worth proceeding with the work. The triggering event is a suspension of work of 2 months or more	8.11.1.3	8.11.1.3	6.10.1.3
Similar provisions are made for either party to terminate the Contractor's employment in the case of damage to existing structures	6.14	6.14	5.7
In the past, these provisions have raised strong legal objections, because they meant that a Contractor could terminate its own employment in the event of, say, a fire, which was caused by its own negligence. Amended wording now precludes this action and, unless the Employer is negligent, limits recovery to direct costs	8.11.2	8.11.2	6.10.2

Insurance cover for terrorist acts for both SBC/Q2016 and DB2016 is dealt with separately in clauses 6.10–6.11 (there are no corresponding provisions in MWD2016). Broadly, if cover is available, it is to be obtained in the normal way (e.g. under option A by the Contractor). Should premiums be increased during the contract period, the addition will be added to the contract sum (i.e. paid by the Employer). Should cover be withdrawn, the Employer has the option to terminate the employment of the Contractor or to continue at its own risk (paying for any damage repair as if it were a variation).

INSURING PROFESSIONAL LIABILITIES (PROFESSIONAL INDEMNITY INSURANCE)

Consultants, including architects, building surveyors, engineers, quantity surveyors and services engineers are at risk from claims for negligent performance of their duties. Thus, they should carry professional indemnity insurance (PII). This also applies to a contractor designing the works (for example, under the design and build contract DB2016) or a portion of the works (e.g. a Contractor's Designed Portion in SBC/Q2016 and MWD2016). PII is always on a "claims made" basis. That is, cover is provided for the current period, irrespective of when work was carried out. Further, many insurers will refuse retrospective cover on new business. When the consultant ceases business, it is necessary to carry run-off insurance to allow for claims arising on previous work. This should be for as long as the consultant is at risk. Safe periods vary depending on the type of work undertaken (for some work, practical liability decays a lot quicker than others). A common recommendation is the periods in the Limitation Act – 6 years from date of breach of simple contract (12 years for a deed contract) or commission of a tort, or 15 years as a long stop.

Insurance policies contain a limitation of cover for any one claim (for the Royal Institution of Chartered Surveyors this must be a minimum of £250,000.00 for a small firm) and for all claims in the period (usually one year). There is normally an uninsured excess and this is substantial for some work. In addition to PII, consultants carry public liability and employer's liability insurance to protect against action from third parties and employees.

JCT contracts SBC/Q2016 and DB2016 only mention PII to the extent that it affects Employer/Contractor relationships. The contracts, therefore, provide clauses related to Contractor's PII where the Contractor carries out design work (6.15). The Contract Particulars allow the parties to state:

- Whether PII is required for the Contractor's Design Portion (CDP) (SBC/Q2016) or the Works (DB2016).
- The level of cover for any one occurrence or for all claims in the period.
- The level of cover for pollution events.
- The period of run-off required (6, 12 or an intermediate period of years).

There are provisions for the Employer, or Architect to inspect the insurance documents, but no provision for the Employer to insure in default and adjust the cost. The latter would not be possible for run-off insurance. It is important for the consultants to inspect the Contractor's PII policies prior to agreeing the contract to ensure that the cover is in place and adequate.

OTHER INSURANCE

Statutory insurance   Not mentioned specifically in standard JCT contracts are insurance requirements imposed by statute. These include insurance of employees by employers for injury[3] and insurance of vehicles used on public roads. All participants in building projects have separate liabilities in these respects.

Third party insurance for employer, consultants and "Employer's Persons"
Neither is mention made of third party public liability insurance for consultants or the Employer when visiting the building site or in other everyday work. If the Employer expects to visit the site or have work carried out under the provisions in JCT contracts for "Employer's Persons", it should make sure that it and its agents carry their own public liability insurance. The Employer and agents should also have insurance covering injury to their own employees, damage to property and, for parties carrying out physical work, damage to their own work.

Materials, plant, tools and machinery   A contractor will also maintain insurance for materials on or off site, plant, tools and machinery. As a major loss related to these items could affect the performance of the contract, it is in the Employer's interest to require (in Preliminaries clauses in the tender/contract documents) that appropriate insurance be taken out. In practice, a contractor will probably maintain an All Risks policy covering all the above risks subject to acceptance and endorsement with the project being undertaken.

Employer business risk insurance   On occasion, an uninsured risk might arise, for example, where a fire damages the Employer's premises and delay and indirect costs are incurred (e.g. loss of liquidated damages, extra decanting costs etc.). The Employer should consider taking out separate business risk insurance. This will protect the employer, but not consultants, or contractors. Business risk insurers will pursue these, if appropriate, using subrogation rights (see below) in their policy with the Employer.

Insurance on hand-over   One point where the Employer has to be particularly careful to arrange insurance is on hand-over of the building on Practical Completion of the Works or a Section, or Partial Possession as allowed for in JCT contracts. Liability for the contractor to insure the works (or portions of the works) ends at these hand-over points and consultants should advise the Employer of the impending liability for the new/refurbished building and contents. Where early use of a new building by the Employer is agreed, the Contractor will remain responsible for insuring the works and should obtain agreement from its own insurers. However, the Employer's staff and visitors will not be insured as insurers assume that a building under construction will only be occupied by professional workpeople. Also, if early use involves bringing in contents, these will not be insured.

Subrogation

Insurers will normally include powers of "subrogation" in their contracts. This allows the insurer to use the name of the insured in taking action against any party thought to be liable for damage. As a second party to a contract (e.g. Contractor, Sub-contractor or Employer) will often have caused a loss through negligence, using subrogation powers will defeat the purpose of the insurance – to cover the works, existing building, contents etc. To avoid this, JCT contracts all require insurance to include the Employer, Contractor and all Subcontractors as insured parties. This is done by requiring the insurance to be in their joint names. As the insurer cannot sue the parties that it is insuring, this defeats the subrogation provision. An alternative for subcontractors is to require that insurers waive the right of subrogation for any subcontractor (SBC/Q2016, 6.9.1.2, DB2016, 6.9.1.2). This provision is not written into MWD2016.

One exception to the joint names requirement is in MWD2016, clause 5.4C. There are no specific insurance requirements and the Employer may simply maintain own name insurance for existing structures and contents. This provision avoids the cumbersome requirement to take out special joint names cover for small works in an existing property. However, in the event of a loss to the structure caused by the negligence of the contractor, the insurer may use subrogation powers to pursue the latter.

Amount of insurance, under-insurance and averaging

It is important that cover provided is for an adequate amount. Buildings insurance premiums are based on full value. That is, it is not normally possible to insure for just the likely loss in the event of a claim. If a loss arises, where the building is under-insured, the insurer will average the claim – that is, reduce the claim proportionately to the amount of under-insurance. The amount of insurance required is calculated on the basis indicated below.

Type of risk	Method of calculation	Basis of risk
New building work (and work in existing buildings)	Contract Sum plus demolition/clearing costs and professional fees	All risks with exceptions

Existing building	Re-building costs plus demolition/clearing costs and professional fees	Named risks (in JCT contracts, called **Specified Perils**)
Contents	As new replacement cost, or value less depreciation	Named risks
Plant and machinery	Value less depreciation	Named risks

Loss adjustment

In the event of a claim, the insurance company will often wish to assess the amount of the loss themselves. They usually appoint loss adjusters to settle on their behalf. For construction and building claims, the loss adjusters are usually quantity or building surveyors with specialist training in insurance. They will be involved in assessing the amount of insurance, the extent of the loss and the accuracy of accounts, estimates for rebuilding etc.

BONDS AND GUARANTEES

Performance bonds

Bonds are a special kind of insurance usually provided by a bank or surety company. They are a guarantee to pay a certain amount on the occurrence of a defined event. The most familiar type of bond used in the construction industry is the performance bond, which guarantees to pay the cost of validly terminating the employment of a contractor under the terms of the contract being used. As termination is usually invoked in situations of contractor insolvency and as insolvency is relatively common in the construction industry, the cost of such a bond is high. For this reason, it is usual to limit the value of the bond to a small proportion of the total contract sum – for example the limit of the bond might be set at 10% of the contract sum. In many instances, the cost of termination, when delay costs and the additional cost of another contractor finishing work is taken into account, is greater than the limit of the bond. However, for performance bonds, the loss has to be assessed and agreed with loss adjusters acting for the surety company.

Unconditional on-demand bonds

Unlike performance bonds, on-demand bonds involve an agreement to pay the whole value of the bond on the occurrence of an event, without showing fault. Therefore, a bond could potentially be called for mere lateness in performance or failures in technical performance caused by parties other than the contractor. These bonds may be used as a threat to prompt additional work. They are sometimes used worldwide by government agencies and are very onerous to the contractor, not the least because the bank providing the bond will treat it as an unlimited obligation, capable of being demanded at any time and for which adequate funds to meet the cost must be available from the contractor. The high risk of providing such a bond makes them very much more expensive than a performance bond.

Conditional on-demand bonds

Conditional on-demand bonds also involve an agreement to pay the whole value of the bond, but the prompting event, including the identifying default of the contractor, will be specified. There will also be a defined mechanism for calling the bond including, perhaps, opportunities for the contractor to rectify underlying issues in a short time. The fact that the full value of the bond is payable on calling also makes this type of bond expensive.

Parent company guarantees

A parent company guarantee does not involve a third party financial organisation but involves a group or holding company guaranteeing the performance of a subsidiary. Thus, should the group wish to terminate the activities of a subsidiary construction company, it will guarantee that it will still perform the obligations of the latter as set out in the contract. Although less certain of payment than a bond, parent company guarantees do not involve payment of any premium, making them an appropriate mechanism for obtaining some security of performance in the right circumstances.

Advance payment bonds

The possibility of advanced payment is covered in the chapter covering interim payment and is appropriate for some types of work – for example for specially ordered pre-fabricated modules or components. Providing advance payment does, however, put the Employer at risk should the Contractor fail to repay the advance payment. JCT contracts envisage that advance payment will be coupled with a conditional on-demand bond (provided by a third party bank or surety company) for the full amount, reducing as the payment is re-paid in interim certificates.

Retention and off-site payment bonds

The traditional method of ensuring that snagging defects are corrected before the end of a contract is to retain a small amount from interim payments. This retention amount, although usually only 3% or 5% of the certified value, is a large figure in relation to the contractor's profit and working capital. The alternative of allowing the contractor to arrange a retention bond, guaranteeing that it will not default in correcting defects, is allowed for in JCT contracts. The cost of the bond is added to the value of the contract but is presumably offset by the lower financing costs to the contractor, which finds its way back to the Employer in lower tenders.

With the increasing mechanisation of building work, there is a continuing move from site to off-site assembly and pressure to make interim payment for goods and work-in-progress not yet on site. Making such payment puts the Employer at risk and provision is now made in JCT contracts for off-site payment bonds. These are only likely to be used where it is envisaged from the outset that substantial parts of the building are to be assembled partially off site rather than brought to site as elements for on-site assembly.

Limitations of bonds

Bonds are normally "recourse guarantees" – that is, the bond provider, whether bank or surety company, will seek to recover the cost of having to pay out on the bond from the contractor. This means that it will be interested in the financial stability of the contractor and will only provide a bond if there are sufficient uncharged assets available to allow recovery in the event of a loss. Employing bonds, therefore, limits tender lists to substantial companies capable of showing sufficient solvency to support their cost. This is not problematic where there are only a few clients with a requirement for bonded performance but will restrict most contractors if widely demanded. If a balance is sought between competitive performance and security, the use of bonds should be limited and other methods of financial prudence adopted instead. These could include more careful scrutiny of contractors' financial data and performance indicators.

How the contracts operate

Topic	SBC/Q2016	DB2016	MWD2016
Advanced payment bond If there is a requirement for an advanced payment bond set out in the Contract Particulars, then payment is only made if the bond is executed	4.7 Schedule 6 part 1	4.6 Schedule 6 part 1	–

Off-site payment bond If there is a requirement for an off-site payment bond set out in the Contract Particulars, for "uniquely identified" off-site items, then payment is only made if the bond is executed to the amount mentioned	4.16.4 Schedule 6 part 2	4.15.4 Schedule 6 part 2	–
There is an expectation that there will be a requirement in the Contract Particulars for a bond for off-site items which are not "uniquely identified". Payment is only made if the bond is executed to the amount mentioned	4.16.5 Schedule 6 part 2	4.15.5 Schedule 6 part 2	–
Retention bond If there is a requirement for a retention bond set out in the Contract Particulars and this is executed, then payment is made without deduction of retention	4.18.1–5 Schedule 6 part 3	4.18.1–5 Schedule 6 part 3	–

There are no provisions for bonds in MWD2016 and no provisions covering other types of guarantees.

ISSUES IN INDEMNITY AND INSURANCE

As the amounts involved are very large and insurance companies are liable to pay much larger amounts than at risk for any one claim if a principle can be established, there is an active body of case law surrounding indemnity and insurance issues in construction contracts. Most recent issues have surrounded attempts by insurers to take action against contractors and sub-contractors, where the Employer has been obliged to insure the work, but the loss is caused by the contractor. The issue has largely been about whether the obligation to insure overrides the indemnity the contractor provides in earlier versions of clause 6.2 of SBC/Q2016 and DB2016 (5.2 of MWD2016) or an implied obligation to take care in carrying out work. In the more recent cases (e.g. the Kruger case below), courts have held that it does not and insurers for the Employer have been able to take action against contractors, if an employee/sub-contractor has been negligent.

Clauses 6.2 and 6.3 of SBC/Q2016 and DB2016 and clause 5.2 of MWD2016 now make clear that indemnity for damage to property excludes the Works themselves and the Employer's existing buildings.

The most recent case (4 below) has seen an insurer successfully take action against a consultant for failure to supervise properly, thus allowing a loss to arise. In that case, the insurers could not take action against the contractor because the insurance was in the joint names of the contractor and employer precluding the use of subrogation to sue the contractor.

Duties on consultant architects and project managers are onerous in relation to ensuring that contractors and subcontractors carry out their obligation to insure. If a consultant does not inspect policies, where the contract requires this, they may be liable for losses arising (5 below). The practical imperative is to always ensure that contractually required insurances have been inspected and logged and to ensure that PII insurers are aware of potential liabilities related to supervision duties, especially where the principal parties have joint names insurance in place.

CASES

1. *Surrey Heath Borough Council v Lovell Construction Ltd* (1990) 48BLR108
2. *The National Trust v Haden Young Ltd* (1994) 72BLR1
3. *Kruger Tissue (Industrial) Ltd v Frank Galliers Ltd and Other* (1998) 57ConLR1
4. *Co-operative Retail Services v Taylor Young and Others*, 2000 (See Building, 7/9/01:64)
5. *Pozzolanic Lytag Limited v Bryan Hobson Associates*, 1999, BLR 267 (See RICS Construction Journal, September–October 2016)

NOTES

1. See Civil Liability (Contribution) Act 1978, 1978 Chapter 47.
2. Unfair Contract Terms Act 1977, 1977 Chapter 50.
3. Employers' Liability (Compulsory Insurance) Act 1969, 1969 Chapter 57.

BIBLIOGRAPHY

Fire Protection Association (2016) *Fire Prevention on Construction Sites Joint Code of Practice*, 8th edition, FPA, London, UK

Central Unit on Procurement (1994) *CUP Guidance No. 48 – Bonds and Guarantees*, HM Treasury, London, UK

10

Fluctuations

INTRODUCTION

The position in general law regarding payment for building works is that (assuming a lump-sum contract) increases in the factor costs of construction during the project period are at the risk of the contractor. This would include increases caused by normal inflation and increases brought about by imposition of taxes etc. (unless these were so onerous that they had the effect of frustrating the contract). The contractor has more knowledge about increased costs and the general risks of building than the employer and thus would normally be better placed to price for them. Excluded from this would be increases caused by the action of the employer. Thus, increased costs are payable, if incurred, on variation additions and similar items.

There is, therefore, no need to include in contract forms clauses covering increased costs if the contractor is to bear the risks. However, it is usual to modify the general law for two broad reasons.

1. For long projects, it is difficult for a contractor to predict increased costs. In these circumstances, a premium would be charged to cover the risk of getting the calculations wrong. It might be better for the employer to allow payment of increased costs and thus achieve tighter pricing at tender stage. For many commercial clients, it is also cheaper for them to obtain finance than the contractor and, given that the client always pays, it may be cheaper for it, rather than the contractor, to borrow.

2. Third parties sometimes impose increases well beyond the control of either party. For example, the government might increase duty on goods or employment taxes. There may be no way that a prudent contractor could anticipate such moves.

HOW THE CONTRACTS OPERATE

SBC/Q2016 and DB2016 envisage four broad alternatives for dealing with increased costs (or fluctuations as they are often called – in order to reflect the possibility of decreasing costs). The options are:

1. a fully fixed price contract, with no allowances for increased costs at all
2. a limited fluctuations contract allowing increased costs for changes in taxes etc.
3. a fully fluctuating contract allowing increased costs based on changes in factor costs
4. a "formula fluctuations" contract (allowing increased costs based on nationally indexed changes in building prices)

MWD2016 envisages two of these alternatives – (1) and (2) and all three contracts make provision for other arrangements to be included. The "default" position in all three contracts is that (2) the limited fluctuations alternative will be used if no other is actively chosen. Only this alternative is written into the contracts (as Schedule 7 for SBC/Q2016 and DB2016 and Schedule 2 for MWD2016), with the others available separately from the JCT.

Topic	SBC/Q2016	DB2016	MWD2016
Fully fixed price contract MWD2016 allows for a fully fixed price contract, but with some limited exceptions. These all relate to additional costs brought about by the actions of the Employer or agents. It would be difficult to deny the Contractor fluctuations costs on these amounts. The provision is intended for very small projects, where changes in taxes imposed by public authorities (see below) and beyond the control of the Contractor are unlikely to be a major concern	–	–	4.9
Exclusions to the fixed price provision Increased costs on variations are excluded as are loss and expense payments associated with variations, or other changes in the works	–	–	3.6

Amounts set against provisional sums are excluded	–	–	3.7
Costs and expenses related to validly suspending the works by the Contractor for non-payment are excluded	–	–	4.7
SBC/Q2016 and DB2016 also allow for a fully fixed price contract, but without expressly detailing exceptions. However, allowances for increased costs where the Employer or Architect has effectively caused the increase can be made elsewhere – for example in ordering variations, the valuation of variations clause (5.6) or loss and expense clause (4.20) can include any increased costs for the item The effect of a fully fixed price contract is, as with MWD2016, to require the contractor to take the risk of fluctuations beyond the control of either party – in particular the imposition of taxes and duties	Contract Particulars to 4.3, 4.14	Contract Particulars to 4.2, 4.12, 4.13	–
Limited fluctuations option All three contracts allow for this option, which does not permit increased costs for normal builder's work. It does allow limited fluctuations covering increases in the rates of duty payable to public authorities in the form of tax on goods and labour levies etc. As increases/decreases in these costs are wholly beyond the control of the Contractor, it is considered by the drafters of the forms (following a similar approach to dealing with extensions of time for external events) that the Employer should bear the risk	Contract Particulars to 4.3, 4.14 and Schedule 7 Fluctuations Option A	Contract Particulars to 4.2, 4.12, 4.13 and Schedule 7 Fluctuations Option A	Contract Particulars 4.3, 4.8 and Schedule 2

of their occurrence, presumably benefiting in lower tender prices as the Contractor does not have to price the risk. VAT is dealt with separately by the contracts and is not adjusted here The action needed to invoke this alternative is to leave the form un-amended as this is the default alternative			
The method adopted by the contract is to state that the prices in the tender sum are based on rates of duty etc. applying at the "Base Date". The "Base Date" is the date inserted in the Contract Particulars against the definitions clause 1.1 and is often in the month prior to the submission of the tender to reflect when the project would have been priced. Should these rates of duty increase or decrease, then the net increase or decrease will be paid to/deducted from the contractor's payments. The figures are calculated in the same way as detailed below for traditional fluctuations			
Traditional fluctuations option This allows fluctuations on materials, plant, overheads and profit calculated on a factor cost basis. That is the actual costs of materials, labour and plant etc., which are reimbursed as they increase on the project. Action needed to invoke this alternative is to refer to Option B in the Contract Particulars	Contract Particulars to 4.3, 4.14 and Fluctuations Option B	Contract Particulars to 4.2, 4.12, 4.13 and Fluctuations Option B	–
The traditional fluctuations clauses work by allowing increased costs expressly on labour, materials, fuel and			

electricity. Plant, overheads and profit are not expressly included but can be covered by way of a general percentage addition inserted into the Contract Particulars. Dayworks are excluded from the increase, as is VAT (adjusted as a separate transaction)			
Subcontractors are also excluded, but there is a requirement for back-to-back provisions in subcontracts and the increase is passed on to the Employer			
Fluctuations apply to "Schedule 2" variations quotations and variations on these quotations (SBC/Q2016), but only if a base date has been specified in the quotation itself. They do not apply to contractor's estimates for valuing changes in DB2016 and the Contractor would need to make allowance in the estimate			
Labour is adjusted by tying labour payments to nationally agreed rates and only increasing them in line with national agreements. This is to avoid the chaos and lack of control that would ensue if the actual (and often informal) payments to labour were to be adjusted. If the nationally agreed rate is increased, the increase multiplied by the hours actually worked beyond the date of increase is paid to the contractor. This method has two main problems; it is administratively cumbersome and it is prone to some manipulation (e.g. by overstating the number of persons on site beyond the increase date)			

Materials, fuel and electricity are adjusted by allowing increases in prices fixed at the "Base Date" mentioned above and the actual cost differences between invoices and initial prices are paid. This method has three main problems; it is administratively cumbersome, it is prone to manipulation of initial and invoice prices (by deliberately understating the former or deliberately overstating the latter) and it is prone to manipulation of the quantities used (i.e. invoice quantities have to be matched to measured usage)			
A percentage addition is quoted by the contractor in the Contract Particulars to deal with indirect costs. These will typically include the following: 1 Differences between labour increases on national scales and actual increases envisaged (national rates typically under-recover in an economic boom and vice versa) 2 Additions for dedicated plant, non-dedicated plant and other on-site overheads 3 Additions for off-site overheads 4 Additions for profit			
As calculating traditional fluctuations are cumbersome and time consuming, express provision is made to short-circuit the process. In SBC/Q2016 by the Quantity Surveyor and Contractor directly agreeing amounts and in DB2016, by the Employer and Contractor doing likewise			

Formula fluctuations option See overleaf	Contract Particulars to 4.3, 4.14 and Fluctuations Option C	Contract Particulars to 4.2, 4.12, 4.13 and Fluctuations Option C	–

FORMULA FLUCTUATIONS

Formula fluctuations work by adjusting output prices rather than input factor costs. This is done on the basis of the national average increased costs of 60 categories of work. The categories of work correspond broadly (but not exactly) with the New Rules of Measurement sections and are meant to represent discrete sub-trades within the average project. Each category is indexed to a convenient base year. The base year is updated from time to time, but this will not affect the calculations as long as it is not changed during a particular project. Increases in the average costs of labour, materials and plant within each category are averaged nationally and indexed once a month to give "current indices". The formula works by comparing the index of a particular category at the date of valuation with the same index as it was at the time of tendering for the work (i.e. at the base date).

$$\text{Category increase} = (\text{CMI} - \text{BMI})/\text{BMI} \times \text{NVWD}$$

For some specialist works, there are two indices, one for labour and the other for materials. For these works, the formula is:

$$\text{Category increase} = ((\text{LCMI} - \text{LBMI})/\text{LBMI}) \times \text{NVLD} + ((\text{MCMI} - \text{MBMI})/\text{MBMI}) \times \text{NVMD}$$

KEY	CMI	Current month index
	BMI	Base month index
	NVWD	Net value of work done
	LCMI	Labour current month index
	LBMI	Labour base month index

	NVLD	Net value of labour work done
	MCMI	Materials current month index
	MBMI	Materials base month index
	NVMD	Net value of materials done

The net value of labour and materials in the specialist indices are split on a standard basis agreed at the contracting stage.

The procedure adopted for formula fluctuations is to split the bills of quantities into the 60 categories at the tender stage and fit the builder's prices into the categories. As work is expended against the categories, the increased costs are calculated using one of the above formulae for each category and the net increase is added to the valuation each month. For work not fitting into any category (known as the "balance of adjustable work"), increases are proportionate to the average increases for all other work. Detailed procedures for applying formula fluctuations are published by the JCT[1] and are referred to in SBC/Q2016 and DB2016.

The main advantages of the formula method of calculating increased costs are:

1. Administratively quick and amenable to computer calculation
2. Difficult to manipulate

The main disadvantages:

1. Does not reflect the actual costs incurred on any one site. In particular the following distortions may occur:
   (a) In a boom, the formula will cause under-recovery in the most buoyant geographical area where costs are rising fastest.
   (b) In a recession, the formula will cause over-recovery in the most affected geographical area where costs are falling fastest.
   (c) The opposite affects will be felt in regions away from the centre of boom/recession.
2. Requires detailed bills of quantities so cannot work with many forms of contract.

NOTE

1. JCT (2017) Formula Rules 2016, Sweet & Maxwell, London.

11

Maintaining quality

INTRODUCTION

Both the traditional forms of contract SBC/Q2016 and MWD2016 are quality-driven contracts. That is, they envisage that separate consultant designers will be appointed to produce detailed designs with a contractor carrying out the work strictly in accordance with the designs. The designers not only produce the detailed design but also perform supervisory functions under the contract, including authorising interim and final payment, practical completion of the work and final certification that the works are complete as intended. It is largely through payment and completion certificates that the quality of work is controlled.

DB2016 has many characteristics in common with the two traditional contracts, but is more flexible, with an emphasis on speed of construction and cost rather than quality. The design is therefore not specified by the Employer or is specified in less detail, with the Contractor providing scope as well as detailed design in many instances. Supervisory consultants are not envisaged with this form and there is no reference to certificates. The impartial flavour of "certificates" is replaced by "statements" produced either by Employer or Contractor depending on the circumstances. Interim and final certificates become interim and final "payments". Other than this, the wording of this contract is similar to the other two forms and the practical approach of the construction team is likely to be similar. Any Employer's Agent will operate payment and statement duties in much the same way as an architect issuing certificates but will not (at least expressly) be acting as an impartial certifier.

SBC/Q2016 and DB2016 also make provision for testing work. Tests may be planned at the outset and written into the tender/contract documents or may be requested by the Architect (SBC/Q2016) or Employer (DB2016) at the time of construction. If the tests fail, or are allowed for in the contract documents, then their cost falls on the contractor. If they pass and are not allowed for in the contract documents, then their cost is added to the contract sum. Problems occasionally arise where planned tests fail, prompting more costly tests of installed work, which pass. The Architect's or Employer's view is likely to be that, but for the failure of the planned tests, the tests of installed work would not be required and the cost should fall on the Contractor. The Contractor's view is likely to be that the installed work has passed the tests and it should be paid!

To manage this conflict of view, the two contracts both make provision for a Code of Practice in relation to unplanned further tests (Schedule 4). The Code basically advises the parties to take a considered approach to the extent, type and method of testing reasonably required by the Architect (SBC/Q2016) or Employer (DB2016). Provided that instructions for testing are reasonable (in the light of the Code), the cost is born by the Contractor, whether the tests pass or fail. However, should the tests pass, the Contractor may be able to claim an extension of time for any delay caused by the testing. Although the minor works contract MWD2016 does not specifically make allowance for testing, tests can be made a contractual requirement by writing them into any specification or schedule in the contract documents.

Compliance is obtained in SBC/Q2016 and DB2016, by the issue of instructions for the removal of work. In SBC/Q2016 there is a provision to accept work at a reduced cost and in both SBC/Q2016 and DB2016 there is provision to change required work to resolve the non-compliance at no cost to the Employer. MWD2016 does not make specific separate provision for dealing with removal of non-compliant work, but the Architect may issue instructions and insist they are carried out. The instructions could cover matters related to such work. Compliance is further backed up by general sanctions on failure of the Contractor to carry out an instruction. Others can be brought in to carry out the work and their cost charged back to the Contractor. In extreme circumstances (in SBC/Q2016 and DB2016), the employment of the Contractor can be terminated, again with provision to charge the cost of getting others to complete to the Contractor.

The personalities of the design and construction team may also be important in achieving desired quality and these are controlled to some extent by all three contracts. Apart from the appointment of named consultant Architect, Contract Administrator, or Employer's Agent, other individuals are mentioned. In the negative, in SBC/Q2016 and MWD2016, the Architect may exclude individuals from the works. This should not be done unreasonably but could include people from whom the Architect has had poor work in the past. In all three contracts, there is a requirement for the Contractor to have a competent person in charge, with express responsibility to take instructions from the Architect (SBC/Q2016, MWD2016) or Employer (DB2016). For larger projects, this individual may be professionally qualified and, whether qualified or not, is likely to have a good knowledge of building construction and quality requirements. All three contracts encourage the employment of registered cardholders under the Construction

Skills Certification Scheme[1] or equivalent. This scheme promotes skill standards for the building trades.

SBC/Q2016 makes further specific provision for a "clerk of works". This individual is employed by the Employer but is, for practical purposes, under the direction of the Architect. His or her main task is to act as a site inspector in ensuring that work is up to standard. In practice, this task has evolved into two duties, inspecting and approving work on behalf of the Architect and issuing instructions in relation to non-compliant work or changes to work brought about by the exigencies of the site. One of the advantages to the Contractor of a clerk of works is having a representative of the Architect on hand during dynamic stages of a project, for example, where structural work is carried out and will be rapidly buried or covered over. Approval by a clerk of works can considerably speed progress. In relation to the instruction duty, SBC/Q2016 does not formally allow for clerk of works instructions but does acknowledge that they will be issued and requires confirmation from the Architect within two working days.

At the corporate level, all three forms of contract influence quality by controlling subcontracting. Assigning the whole of the contract by Employer or Contractor is prohibited without the consent of the other party. Permission is required of the Architect (SBC/Q2016, MWD2016) or Employer (DB2016) for subcontracting of the whole or part of the works. In addition, with DB2016 (and the other two contracts for any Contractor Designed Portion), permission is required to subcontract the design. SBC/Q2016 and DB2016 allow for named subcontractors, ensuring that firms known to and approved by the Architect/Employer are engaged for elements of the project. SBC/Q2016 also allows for Named Specialists, allowing the Architect to specify a particular specialist subcontractor for sections of the work.

It is usual for new buildings to contain minor defects when completed (termed "snagging" defects) and normally a contractor would be given an opportunity to put these right for a short period of time (in JCT contracts, termed the "rectification period"). For the traditional contracts SBC/Q2016 and MWD2016, the Architect will, at Practical Completion, produce a list of minor defects and update this towards the end of the period (pointing them out at the same time to the Contractor's representative). He/she will return to ensure they have been corrected. Although certifying that the corrective work has been done, this acknowledgement does not absolve the Contractor for responsibility for correcting defects later. For later defects the Employer may go back to the Contractor, or get others in to do the work and try to charge the cost back to the Contractor. DB2016 makes similar provision, but the Employer produces the snagging list and issues a notice, rather than certificate, indicating that defects have been corrected. Certifying or giving notice, as appropriate, is an indicator that the works are finally complete and that final payment may be made, subject to producing the accounts.

HOW THE CONTRACTS OPERATE

Quality requirements			
Topic	SBC/ Q2016	DB2016	MWD2016
Comply with the contract documents The quality requirements are as set out in the contract documents and the Contractor is required to carry out the works in accordance with these requirements **Comply with statutory requirements** These would include, where relevant, compliance with building regulations, but for traditional contracts using SBC/Q2016 and MWD2016, design-related aspects of building regulations would be dealt with by the Architect (Except for any CDP)	2.1	2.1.1	2.1.1
Fill in details of design and specification In DB2016 and for Contractor Designed Portions of the other two contracts, the Contractor is also required to fill in details of design not expressly covered in the contract documents	2.2.1	2.1.1	2.1.1
Work to be up to standard Both SBC/Q2016 and DB2016 expressly state that materials, goods and workmanship are to be of the standards described in the contract documents	2.3.1–2	2.2.1–2	–
Provision of samples DB2016 states that samples are to be provided, if required by the contract documents. SBC/ Q2016 and MWD2016 do not expressly state this, but not providing samples required in contract documents would be non-compliance with the general requirement to comply with the documents	–	2.2.3	–

Work to the subjective standards of the Architect SBC/Q2016 and MWD2016 allow the Architect to express satisfaction or otherwise with the quality of materials, goods and workmanship. Whilst this may be a reasonable approach, particularly zealous individuals may set the bar too high! The contracts, therefore, emphasise that satisfaction must be reasonable. Reasonableness would be open to external assessment (perhaps at adjudication) against the standard of other Architects in similar situations	2.3.3	–	2.2.1
Work not to any express standards SBC/Q2016 and MWD2016 expressly state that materials, goods and workmanship which are neither to an objective standard in contract documents nor to the satisfaction of the architect are to be of a standard appropriate to the works	2.3.3	–	2.2.1
Proof SBC/Q2016 and DB2016 require that the Contractor provide proof that materials and goods are up to standard The omission of "workmanship" from this requirement indicates that this requirement is not the same as asking for tests of finished work. Although there is no similar requirement in MWD2016, provisions could be written into contract documents, such as a general specification or schedules of work	2.3.4	2.2.4	–
Payment			
Interim payment Interim payment is made for work *properly* executed. In other words, the Architect (SBC/Q2016, MWD2016) or Employer (DB2016) deducts the cost of work improperly executed, until it is put right, accepted at a discount or at no additional cost	4.14.1.1	4.13.1.1	4.3.1

DB2016 makes provision for stage payment and where this applies (Alternative A of the Gross Valuation Ascertainment), there is no mention of work being properly executed. However, the Employer could argue that a stage has not been reached if the work therein contains defects and may be able to withhold payment for the whole stage until defects are corrected	–	4.12.1.1	–
Payment on practical completion For all three contracts, when the project has reached practical completion, the Contractor is entitled to payment of part of the retention, release from the requirement to insure the works and release from the obligation to pay liquidated damages thereafter. For SBC/Q2016 and MWD2016, the Architect certifies when practical completion has been achieved. For DB2016, this state of affairs simply has to be achieved, for the Employer to issue a Practical Completion Statement. The Architect, Employer or Employer's Agent could argue that defects in the work mean that Practical Completion has not been achieved and refuse to issue the Certificate/Statement. Unresolved disputes on the matter might be referred to mediation or adjudication	2.30	2.27	2.10
Provisions related to interim payments regulate the amount of retention to be (effectively) released on achieving Practical Completion	4.19.2	4.18.2	4.3
Final payment All three contracts state that the contract sum will be adjusted by deducting amounts related to failure to carry out an Architect's (SBC/Q2016) or Employer's (DB2016) instruction and having to get others in to correct the work	4.3.6 3.11	4.2.6 3.6	4.8.2.1 3.5
SBC/Q2016 and MWD2016 also allow the Architect to accept work not properly executed at a discount	4.3.6 3.18.2	–	4.8.2.1 2.11

The cost of tests is added to the contract sum, provided the tests pass and have not previously been provided for in the contract documents	4.3.6 3.17	4.2.6 3.12	–
MWD2016 makes no separate mention of how the contract sum is to be adjusted for the cost of tests. However, a reasonable approach would be to follow the SBC/Q2016 approach on the basis that it is best practice guidance	–	–	

Testing

The Architect (JCT/Q2016) or Employer (DB2016) may order tests. If they pass, the cost is added to the contract sum. If they fail or are allowed for in the contract documents, there is no financial adjustment	3.17	3.12	–

Compliance

If work is not in accordance with the Contract, the Architect (SBC/Q2016) or Employer (DB2016) may			
• Issue instructions for removal	3.18.1	3.13.1	–
• Accept work at a discount	3.18.2	–	
• Vary (SBC/Q2016) or Change (DB2016) work to make it acceptable, but without additional payment or time	3.18.3	3.13.2	–
• Order further tests (following the Code of Practice mentioned above), without further cost. An extension of time may be granted if tests are passed but a delay ensues	3.18.4	3.13.3	–
If workmanship is not in accordance with the Contract, the Architect (SBC/Q2016) or Employer (DB2016) may issue instructions, but without additional payment or time	3.19	3.14	–

There are no conditions in MWD2016 corresponding to the above testing and compliance clauses, but the Architect has wide power to issue instructions and authorise payment and could get the same result using these more general powers and, perhaps, following the SBC/Q2016 approach on the basis that it is best practice guidance			
Enforcing instructions			
In all three contracts, there are provisions requiring compliance with instructions and allowing the Employer to get others to carry out remedial work and charge this back to the Contractor	3.11	3.6	3.5
SBC/Q2016 and DB2016 allow the Employer to terminate the employment of the Contractor should it refuse to remove non-conforming work, materials or goods (provided that this failure materially affects the work). MWD2016 does not expressly allow this, but others could remove the work at the Contractor's cost and, for smaller works, this remedy is likely to be sufficient	8.4.1.3	8.4.1.3	–
Personalities			
Named consultants, with defined duties under the contract. SBC/Q2016 and MWD2016 envisage an architect as the primary consultant (or contract administrator, if not a registered architect). It is this individual who has primary authority for ensuring that quality standards are met In DB2016, the Employer directly undertakes many of the duties assigned to an architect in the other two contracts. However, this contract also envisages the appointment of an Employer's Agent, who is given full authority to act as the Employer	Article 3	Article 3	Article 3

| Workpeople
All three contracts require the Contractor to encourage all workpeople under its control to be registered cardholders under the Construction Skills Certification Scheme (CSCS) or equivalent	2.3.5	2.2.5	2.2.2
SBC/Q2016 and MWD2016 allow the Architect to exclude individuals from the works. There are no equivalent provisions in DB2016	3.21	–	3.8
Contractor supervision
Requirement for a competent person on site	3.2	3.2	3.2
Clerk of works	3.4	–	–
Assignment and subcontracting
Assignment prohibition	7.1	7.1	3.1
Permission required for subcontracting whole or part of the works	3.7.1	3.3.1	3.3.1
Permission required for subcontracting the design	3.7.2	3.3.2	3.3.1
Named subcontractors	3.8	Sch 2 Part 1 Para 1	–
Named specialists	Sch 8 Para 9		
Snagging			
Production of the snagging list during the rectification period	2.38.1	2.35.1	2.11
Putting right the defects (or accepting them at a reduced cost)	2.38	2.35	2.11
Certificate or notice of making good defects	2.39	2.36	2.12

PRACTICE

Pre-contract

All three contracts maintain quality by setting out the quality requirements in the contract documents and then by making provisions for ensuring that the contractor sticks to these. SBC/Q2016 and MWD2016 envisage that the enforcing authority will be an architect, but DB2016 expects the Employer (or an agent) to fulfil this requirement. Where bills of quantities are involved (SBC/Q2016), the New Rules of Measurement (NRM2) ensure that all significant items are included, with expressly defined standards. The pre-contract process involves the architect in producing detailed drawings from which the quantity surveyor will measure items to include in the bills. Some provision may be made for sampling in the bills – for example it is common to ask for samples of bricks and tiles to be used on a traditional project, perhaps with a sample panel being built before ordering the bulk quantities. The Contractor will price for providing the samples and the cost is included in the contract sum. Tests are often envisaged at the outset and express provisions are written into the bills for testing. For example, cube tests and slump tests[2] may be envisaged for concrete to be provided in the foundations and frame of a building. Again, as the tests are priced by the contractor and the cost is included in the contract sum, there will be no additional payment if the tests are executed as planned.

For smaller projects, MWD2016 follows a similar approach, but the documents are less carefully organised. There is no equivalent to NRM2 and documents may be simple drawings and specification. Some projects using the form can be quite complex and contain general specifications, schedules of work and detailed preliminaries. Provision may be made in the last for samples and testing, but, if not, these will be at additional cost and, possibly, at additional time if the work is delayed. There may also be considerable gaps in the documents leaving both more to the discretion of the contractor and/or to the approval of the architect. For the former, a reasonable standard (viewed objectively) would need to be provided, but it is common for an architect to require at least all visible elements to be either to an objective standard or to his/her approval.

DB2016 envisages that the employer will set up requirements at the outset, with the contractor supporting these with detailed proposals. The quality requirements may, therefore, be very limited and, unless supported by detail from the contractor, left to implied terms related to fitness for purpose and merchantable quality. For projects with limited quality requirements, the building regulations provide a significant control as DB2016 requires that the contractor complies with statutory requirements. In these circumstances, the employer may get a building to the building regulations minimum standard and there may be significant shortcomings in relation to, for example, services and finishings. Key roles of the Employer's Agent are to ensure that requirements accurately reflec the employer's needs and to ensure that contractor's proposals are sufficiently detailed to cover quality requirements.

Post-contract

Traditional contracts using SBC/Q2016 and MWD2016 have a well-worn process for inspecting, testing, correcting and approving work. Architectural supervision will be intermittent, perhaps involving a visit to site on a weekly basis supplemented by visits at key events (e.g. exposure of foundations to an existing building). Contract documents often supplement the requirements in the forms, by requiring the Contractor to give notice when key events are achieved or by imposing general requirements that inspection will be invited before work is covered up. Where a clerk of works is employed, inspection can be, more or less, continuous and this can speed key decisions and operations of both architect and contractor. Quality will also be considered more formally at site meetings, where problems may be aired by the design and construction team. Meeting frequencies will vary but often are weekly, fortnightly or monthly. The basic approach envisaged by the forms is that inspection will be sufficiently regular to avoid nonconforming work getting too far before it is corrected. Where work is wrong, often instructions to remove or correct it will be made verbally and, if there is no dispute, no formal instruction will be issued. It is only when there is a difference in view between architect and contractor that instructions need to be issued and the formal machinery of the forms operated.

Where tests are written into the contract documents, these will be ordered to suit the programme of work. For example, tests on concrete will be ordered up to a month before use if 28-day strengths are required from the test cubes. Occasionally tests are required, which are unplanned, and these should also fit into the programme. Their cost will be added to interim and final payments, provided they are satisfactory. If not, there will be no payment and further tests may be ordered. It is here that the Code of Practice for dealing with testing becomes relevant as there are frequently arguments related to the necessity and relevance of the testing. An example of a further testing problem relates to the cube tests on concrete mentioned above. By the time that test results indicating failure are known, work may have progressed to construction and the only way to ensure that finished work complies with concrete strength requirements is to cut cores from the concrete element and test these. Cutting cores can be an expensive process and there may be an argument over who pays. The Code of Practice now makes clear that, provided the tests are reasonable in the context of the testing requirement, the contractor bears the cost. However, it could be envisaged that, even with a fairly clear scenario such as this, the testing may be unreasonable. For example, asking for core tests for mass concrete filling or trench-fill foundations may be unwarranted for a slightly under performing test.

It is not always necessary to correct defective work and problems can be avoided by working out an alternative solution that is satisfactory, but not as allowed for in the contract documents. If the architect allows this and it is more expensive than the work as specified, the extra cost is borne by the contractor and no adjustment is made to the contract sum. It is also possible that the employer would accept work, which may be adequate, but not as specified. In this instance, a reduction in cost may be forthcoming, the level of which is subject to negotiation in the context of the extent of the concession.

With SBC/Q2016, the Quantity Surveyor will be valuing work for interim payment and passing the valuation to the architect to certify. It is at this point that the architect will make any deduction for uncorrected defective work. In practice, the Quantity Surveyor may have a view on the adequacy of work and may know that a deduction should be made. In this case, the Quantity Surveyor will advise the architect but leave the adjustment to the latter. This avoids the Quantity Surveyor attracting liability for quality. As the deduction is made on the certificate and not on payment (by the Employer), the amount is not a set-off in accordance with the Construction Act provisions[3]. By the time that final payment is due, issues related to accepting non-conforming work at a discount, getting others in to correct work and charging back to the contractor will have been settled and are merely reflected in the accounts.

It is rare for issues related to quality to become so entrenched that others need to be brought in to correct defects or that action is taken to terminate the contractor's employment. Uncorrected defective work is, in fact, often associated with inability due to pending or actual insolvency and the termination action is for the latter. Occasionally, where large differences in view prevail, quality becomes a matter for formal dispute resolution, initially at mediation or adjudication.

In a well-run project, quality issues are, therefore, dealt with as they arise. Snagging is an expected process dealt with amicably between architect and contractor's site manager. Items would normally relate to finishings or services and not substantial lapses in quality. Whether the project is sufficiently completed to enter the rectification period is governed by the issue of the Certificate of Practical Completion. Major defective work indicates that the project is not yet completed, will allow the architect to withhold the Certificate and will expose the contractor to liquidated damages and continuing retention.

Procedures with DB2016 are much the same as the traditional forms, but the standards against which quality requirements are drawn are much more variable. There may also be far less expert consultant supervision with this form, which is likely to be problematic as effectively operating DB2016 provisions requires a detailed knowledge of construction technology. Where an Employer's Agent is used, supervision of quality as envisaged in the contract, requires that this individual be architecturally qualified, or has access to relevant further advice. As most agents are drawn from the established professions and are familiar with the traditional forms, details in Employer's Requirements and Contractor's Proposals related to quality are likely to be similar to their traditional counterpart and the contract will be operated in much the same way as traditional.

NOTES

1. For details of the scheme, see http://www.cscs.uk.com/.
2. Cube tests involve forming a small sample cube of concrete, leaving to cure and then measuring its crushing strength. Slump tests give an indication of the quality and water content of an upturned unset cone of concrete, by measuring the extent that the cone slumps when the cone former is removed.
3. See Chapter 14 for details of the Construction Act.

12

An introduction to dispute resolution in construction

THE ALTERNATIVES

With construction being such a complex assembly process, using many independent agents, it is not surprising that differences sometimes arise. When they do, there are two main ways of attempting to deal with them, internally by direct negotiation between the disputants, perhaps by reference to an involved, but separate third party, or externally by reference to an independent third party. Direct negotiation is always available to the parties, but for many building contracts there is also a contract administrator involved in designing the building and certifying payment and satisfaction with the work. The act of certifying itself demands a degree of impartiality and both the Employer and Contractor can look to the certifier to be impartial. There is, however, a key difference between certification and dispute resolution, in that for the former a dispute has not expressly arisen – the certifier is simply giving a fair and objective opinion on the matter in question. Negotiations may have preceded the issue of a certificate and a certificate may later be corrected if it proves wrong, but if it is still not satisfactory, then a further method has to be found to achieve a resolution.

Dispute resolution, therefore, starts from the point at which negotiation and impartial certification within the normal contract machinery cannot achieve agreement. An independent third party needs to be involved. Having to appoint such an individual tends to polarise matters immediately because the cost of the appointment has to be borne by one or both protagonists. The resolution is, therefore, never better than "win-lose" and often ends up as "lose-lose" situation. If at all possible, disputes should be settled by negotiation within the contract machinery and any external reference should be approached as a last resort. Even then, every attempt should be made to make the resolution consensual, informal and amicable, without developing entrenched attitudes. This will maintain working relationships, save time and keep costs to the minimum. Ranging resolution options broadly in the order of least to most formal, there are three consensual options and four binding options:

- Consensual resolution
  1. Mediation
  2. Conciliation
  3. Neutral evaluation
- Binding resolution
  1. Expert determination
  2. Adjudication
  3. Arbitration
- Litigation

Additionally, many projects now include dispute prevention mechanisms, which may render unnecessary the need for resolution, whether consensual or binding. These mechanisms are considered later.

Consensual resolution

Mediation

Anyone can mediate and mediation can take any form, the success of the procedure simply being evidenced in achieving agreement between the confronting parties. This flexibility arises from the fact that an agreement cannot be imposed on the parties and mediation is really a form of assisted negotiation. However, agreement is likely to be encouraged by certain features common to other forms of dispute resolution. These features include:

- Using a mediator who is respected by the parties. Respect is likely to be evidenced by expertise in the subject matter of the dispute, in construction law, or both.
- A fair and balanced procedure, giving both sides of the dispute a fair hearing.
- Fully engaging in the problem and examining the subject matter.
- Using a mediator with appropriate skills at negotiating towards agreement.
- The use of hearings on neutral territory, where neither party will feel disadvantaged.

Good practice in conducting mediation is contained in several guides, including the **JCT practice note 28** (Mediation on a Building Contract or Sub-Contract Dispute) **ACA** and **ICE Conciliation procedures**. These are summarised in Burkett (2000).[1]

As protagonists often cannot stand the sight of each other by the time they have got even this far, mediation often takes the form of the mediator taking proposals from one party to the other and back. If agreement is achieved, it is prudent to make it into a formal legally binding contract, particularly where one party is making a concession in exchange for no material consideration from the other party. Without consideration, it may be difficult to hold the conceding party to the agreement.

The overriding advantages of mediation are that it is inexpensive, relatively speedy and preserves business relationships. It is also less formal and allows more detailed explanation and examination of each party's position. As negotiations are normally on a without prejudice basis and parties will not be able to use information from the mediation in other action, there should be less defensiveness and/or need for legal representation. This may encourage a more commercial approach to a solution with concessions being made, which in other circumstances would not be forthcoming.

Conciliation

Conciliation is a form of mediation where the conciliator is sufficiently skilled in the subject matter to constructively propose solutions. The conciliator is, therefore, likely to be primarily a technically qualified person, rather than legally qualified. The key advantage of conciliation (over and above mediation) is that, by actively proposing solutions and seeking alternative ways through a problem, the conciliator may be able to bypass it to the benefit of both parties.

Neutral evaluation

This form of determination is a "dry run" trial – sometimes called a "mini-trial". The disputants agree to put the dispute to an expert in advance of taking the matter further. The neutral evaluator will normally be from the legal professions and have considerable experience – perhaps a retired judge or advocate. The process is useful for disputes with significant legal points, but where the law is fairly certain and matters are unlikely to be referred for decisions on appeal. The neutral evaluator should be able to give an accurate indication of outcome at a fraction of the cost of the full trial process. The process would not work well where there are substantial legal points without clear precedents that need decisions in a higher court. It can, however, save considerable cost in most other cases. The hope is that one or both parties sees the merit or hopelessness of their case and use this enlightenment to negotiate a solution.

Binding resolution

Expert determination

Expert determination is legally binding and the requirement to submit to the process may be written into a contract at the outset. However, the agreement to submit to a determination is sometimes made after the dispute has arisen. The binding character of the process means that some formality is necessary, not just desirable as it is in mediation. The expert must treat both parties fairly and be seen to be independent. Expert determination is also a common imbedded feature in some disputes, for example where independent valuers are used in lease agreements to determine rental value and under the Party Wall etc. Act 1996, where the "Third Surveyor" determines disagreement between two party wall surveyors. These disputes generally cover straightforward issues of a technical nature within familiar bounds. Resolution by the expert offers considerable cost savings, particularly when compared with adversarial court action, where experts might be appointed by both sides and still not come to a consensus. Expert determination is not common in building contract disputes but could be used for certain aspects – for example in a dispute on construction quality, agreement might be reached to use a quantity surveyor as an independent expert to determine costs.

Adjudication

"Construction Act Adjudication" has become a feature of the UK construction industry and similar procedures have been developed worldwide. It is largely driven by demands for cheap, quick and straightforward mechanisms to deal with common areas of dispute – non-payment, late payment or disputes on quality, often, but not always, between main and subcontractors. In the United Kingdom, the Housing Grants, Construction and Regeneration Act (1996) (colloquially called the Construction Act) as amended by the Local Democracy, Economic Development and Construction Act (2009) governs "statutory adjudication". The mechanism of the Act (as amended) is to require that provisions for adjudication are written into all building contracts except those with residential occupiers. If adequate provisions are not made in a building contract, then the "Scheme for Construction Contracts" (A Statutory Instrument) will apply, writing standard adjudication terms into the contract! Minimum provisions include that the contract must allow a party to refer a dispute to an adjudicator (either named in the contract or selected from an appropriate professional body), who is required to come to a decision on the dispute within a very short time. This decision will be binding until the contract is finished or abandoned. Thereafter, the parties can agree to accept the decision as final, or use whatever other final resolution mechanism is allowed for in the contract. As adjudication has become such a large and important area of resolution in recent years, it is dealt with separately in Chapter 13.

Arbitration

Although sounding rather like adjudication, arbitration is quite different in that it is dealt with directly by statute (in England and Wales, the Arbitration Act

1996) and may be final and binding on the parties. The parties may agree in a building contract (or subsequently) to use arbitration as the method of settling disputes, but, thereafter, how the arbitrator conducts the arbitration is controlled by the Act. The final and binding nature of arbitration means that the parties should be careful in selecting the method as there is little scope for appeal elsewhere. Arbitration is generally used (rather than litigation) in order to have a referee who is expert in the subject matter of the dispute but that often brings the attendant disadvantage that the arbitrator may be less expert in the law – a major consideration in complex cases!

Litigation

The settlement of legal matters, whether criminal or civil, is conducted at public expense by public officials. However, there is a long-standing acceptance that, within the law, contracting parties may agree how their contract will be run and this extends to dealing with any disputes that may arise between them. Arbitration has, therefore, developed as a private method for resolving contractual disputes. It is not, however, possible to entirely supplant the law courts and, subject to the Arbitration Act, they will intervene if the content of a contract dispute is a public matter. An example of this intervention might be where a dispute involves elements of fraud. As a criminal matter, fraud is always a matter for litigation. Further, should contracting parties have no arbitration agreement in place, or the Arbitration procedure itself is improperly conducted, a dispute may become a matter for the courts.

Many construction disputes involve difficult questions of law and large sums of money and, even within the Court system, are too large to be dealt with by the lower cost courts of first instance. The starting point is often the High Court and, in particular, the Technology and Construction Court. At this level, the costs of the hearing become considerable, often in excess of the matter in dispute – giving rise to the often quoted reason for continuing a dispute – fighting for the costs! The cost and time associated with Court action has led to the Technology and Construction Court itself to encourage fuller examination of disputes through consensual methods before Court action. The Pre-Action Protocol for Construction and Engineering Disputes[2] encourages the parties to use all reasonable alternative methods before bringing the action and this has been a spur towards the use of mediation.

Prevention mechanisms

In larger building and civil engineering projects, it is more common for disputes to arise and this may be anticipated by using prevention mechanisms. Some mechanisms, such as early warning notices, with or without express sanctions, have been written into contracts with a view to heading off problems before they develop into disputes. A further development is the appointment of an individual or panel of individuals to hear disputes as and when they arise. The "disputes adviser" or "disputes board" may be configured to provide non-binding project mediation, where the advisor or board actively contributes conflict management expertise on an ongoing basis. Alternatively, the adviser or board may be configured to comply with the requirements of Construction Act adjudication, with temporarily binding solutions pending review by arbitration or litigation. Advisers and boards allow early pre-emptive resolution of disputes before they become entrenched. They also ensure that an agreed resolution machinery is in place before a dispute has arisen – a much easier proposition than once the parties have quarrelled. This should both make resolution cheaper and help to maintain good relations on the project. One downside to using these mechanisms is the possibility of encouraging argument, given the ready availability of a referee! Another downside, where it is tied to binding Construction Act adjudication, is that the conciliatory attitude towards non-binding advice may be lost to defensiveness and formality.

Binding resolution methods compared

As it is relatively easy to invoke the process during construction, Construction Act adjudication has become the main resolution method in construction disputes, with arbitration or litigation used as an appeal mechanism. Adjudication was set up as a method for rapidly dealing with disputes as they arise and the resolution is speedy, but the result might not be correct. However, as the experience and expertise of adjudicators develops, the quality of decisions should improve and adjudication will produce an increasing number of decisions that the parties feel are not worth challenging. Adjudication does, however, still have some problems in common with arbitration when compared with litigation, in particular that it only deals with disputes between two contracting parties. Many construction disputes involve numerous interlinked issues on separate contracts, all of which have to be separately adjudicated. This is extremely inefficient when compared with the powers of a judge at litigation to join all related disputes into one action.

Comparison of the two main final dispute resolution mechanisms, arbitration and litigation, can be summarised in Table 12.1 overleaf.

Table 12.1 Arbitration and litigation compared

	Arbitration	Litigation
1	**Expertise**	
	Arbitrator can be an expert in subject matter of dispute but may be less knowledgeable of the law. This is mitigated in modern arbitrations, where arbitrators are often dual qualified in law and a technical profession	Judge is expert in law but may require more extensive technical briefing. This is mitigated in modern Court actions where judges specialise in technical areas of the law, such as construction hearings in the Technology and Construction Court
2	**Formality**	
	Procedures may be less formal with private informal venues and less formal dress. The time and place of the hearing is subject to internal agreement between the parties	Process takes place in formal public courts with formal dress including wigs and gowns. The time and place of the hearing is set by the courts schedule
3	**Time and cost**	
	Held out to be more speedy, but, in practice, the extent of examination is likely to be similar to a court hearing. Held out to be less expensive, but this depends on the parties forgoing counsel, expert witnesses etc., which they are unlikely to do if the other side does not and/or the amounts in dispute are large The facilities, including venue, have to be paid for	Waiting times for hearing may be lengthy and **counsel** required for High Court. **Queens Counsel** may be involved for larger cases The premises and facilities are provided at public expense
4	**Confidentiality**	
	Arbitration hearings may be held in private, ensuring confidentiality	Justice must be seen to be done, so the public is admitted as a matter of policy

Table 12.1 Arbitration and litigation compared (*Continued*)

	Arbitration	Litigation
5	**Efficiency**	
	Arbitration is subject to individual arbitration agreements. Related disputes have to be dealt with separately unless all arbitration agreements can be engineered to allow joining of related disputes	Courts can join disputes and deal with them in one hearing
6	**Issues involving crime**	
	Arbitrator cannot deal with issues relating to criminal law, which means, for example, should fraud be involved in the issue the hearing may have to be suspended	Courts have an inherent jurisdiction in all criminal cases and can deal with both civil and criminal matters
7	**Finality of resolution**	
	Under the Arbitration Act 1996, there are very limited rights of appeal even for points of law	In theory, there is a consistent appeal structure from county courts to the House of Lords, but appeals will only be allowed from the High Court for significant points of law

HOW THE CONTRACTS OPERATE

Mediation			
	SBC/Q2016	DB2016	MWD2016
Encouragement to use mediation All three contracts require the parties to "give serious consideration to any request by the other to refer the matter to mediation"	9.1	9.1	7.1

This is subject to either party's right to refer the matter to Adjudication			
Adjudication			
Right to refer matter to Adjudication All three contracts comply with the Construction Act, by giving the parties the right to invoke Adjudication	Article 7	Article 7	Article 6
Procedure for Adjudication All three contracts invoke the Scheme for Construction Contracts as the mechanism by which Adjudication will be operated	9.2	9.2	7.2
Arbitration			
Requirement to refer matter to Arbitration For all three contracts Arbitration must be selected (in the Contract Particulars) for the requirement to refer matters to Arbitration to apply. Matters related to the Inland Revenue Construction Industry tax deduction scheme (CIS) and VAT are excluded from this requirement (as they are dealt with by other tribunals). Disputes related to enforcing decisions on Adjudication are also excluded as this would involve revisiting a temporarily binding decision (but this does not exclude using Arbitration to consider the decision itself at the appropriate time)	Article 8	Article 8	Article 7

Procedure for Arbitration All three contracts refer to the Construction Industry Model Arbitration Rules (CIMAR), which contain detailed procedures for conducting arbitrations	9.3–6	9.3–6	7.3 Schedule 1
All three contracts allow the parties to appeal to the courts on questions of law. If Arbitration is to be final, this provision would need to be amended to make clear that such appeals would not be allowed	9.7	9.7	Schedule 1, para 5
Litigation			
Requirement to refer matter to litigation Litigation applies by default if Arbitration is not selected in the Contract Particulars	Article 9	Article 9	Article 8

There are currently no standard provisions in the three contracts for other methods of dispute resolution.

NOTES

1. See Bibliography to this chapter.
2. Ministry of Justice UK (2017) Pre-Action Protocol for Construction and Engineering Disputes. http://www.justice.gov.uk/courts/procedure-rules/civil/protocol.

BIBLIOGRAPHY

Blackler A, Burkett J (1998) *Association of Consultant Architects (ACA) Model Conciliation Procedure*, Association of Consultant Architects, London UK

Burkett J (2000) *Disputes without Tears – Alternative Methods of Dispute Resolution*, RIBA Publications, London, UK

Grossman A (2009) *Good Practice Guide: Mediation*, RIBA Publishing, London, UK

Institution of Civil Engineers (2012) *ICE Mediation/Conciliation Procedure*, ICE Publishing, London, UK

Joint Contracts Tribunal for the Standard Form of Building Contract (JCT) (1995) *Practice Note 28, Mediation on a Building Contract or Sub-Contract Dispute*, RIBA Publications, London, UK

Joint Contracts Tribunal for the Standard Form of Building Contract (JCT)/Society of Construction Arbitrators (2016) *Construction Industry Model Arbitration Rules (CIMAR)*, Sweet and Maxwell, London, UK

Office of Government Commerce (OGC) (2002) *Dispute Resolution Guidance*, OGC (HM Treasury), Norwich, UK

13

The construction act – adjudication and payment

HOW THE CONSTRUCTION ACT WORKS

The Construction Act is two parts of two Acts of Parliament enacted to deal with several issues. The first, principal Act is the Housing Grants, Construction and Regeneration Act 1996. Housing Grants and Regeneration are irrelevant to construction and the relevant section is Part 2. This part also deals with two separate issues, **statutory adjudication** and **payment**, the objects being to allow speedy and economical resolution of disputes under construction contracts and to ensure contractual agreements to pay (particularly on account) are fair and require prompt payment. The second, subsidiary Act is the Local Democracy, Economic Development and Construction Act 2009. Its purpose is to make some incremental improvements to the operation of the adjudication and payment machinery in the principal Act and to curtail certain abuses in contracts. These include the use of "pay when certified" clauses.

The operation of the act

The easiest way to explain exactly how the Construction Act operates is to cite each relevant section and explain, where necessary. The Local Democracy, Economic Development and Construction Act 2009 maintains the detail of the principal Act, so the clause numbers remain as the principal Act.

HOUSING GRANTS, CONSTRUCTION AND REGENERATION ACT 1996 – PART 2 (THE CONSTRUCTION ACT)[1] AS AMENDED BY THE LOCAL DEMOCRACY, ECONOMIC DEVELOPMENT AND CONSTRUCTION ACT 2009[2]

Statutory adjudication

Section	Subject	Comments
104 (1)	Construction contract	Defines meaning of terms
	a Construction operations b Arranging construction operations c Providing labour for construction operations	Not just construction itself – also includes labour only work and construction management
104 (2)	a Architectural and surveying work b Advice on building, engineering work etc.	Act applies to design and construction consultancy contracts
105 (1)	Construction operations	Defines meaning of terms
	a Construction, dismantling of structures forming part of land b Engineering works etc. c Building services d Cleaning buildings (whilst re-furbishing) e Preparatory work f Painting	Includes most building, engineering, maintenance and demolition work, defined by the exclusions below
105 (2)	Construction operations excludes	
	a Drilling for oil etc. b Mining etc. c Plant, machinery and steelwork in process plants etc. d Off-site fabrication e Artistic works	Relates mainly to work in industries other than construction – (c) being the exception
106 (1)	Contract with residential occupier (excluded)	Does not exclude from the Act subcontract work for residential occupier, nor contract with non-resident landlord

107 (1)	Contract need not be in writing	An amendment from the 2009 Act
108 (1)	Right to refer a dispute arising under the contract for adjudication	Excludes issues (for example) of whether there is a contract in existence, so does not deal with all project matters
	Dispute defined as any difference	The only requirement is that there has to be a difference, so cannot be used for third party determinations etc.
108 (2)	The contract shall include provisions to	Eight points to be incorporated in a construction contract
	<ul><li>Enable a party to give notice at any time</li><li>Provide for appointment of and referral to Adjudicator within 7 days of notice</li><li>Require a decision from the Adjudicator within 28 days unless a longer period of time is agreed "post referral"</li><li>Allow for 14 day extension of decision time with the consent of the referring party</li><li>Require the Adjudicator to act impartially</li><li>Enable the Adjudicator to take the initiative in ascertaining the facts and the law</li></ul>	Six of the eight points are in this section and include a very short time period for a decision, with limited provisions for extending time by agreement. The requirement to act impartially would be implied, but the point allowing the adjudicator to take initiative allows for a more inquisitorial procedure if appropriate
108 (3)	The contract shall provide in writing that adjudicator's decision to be binding until finally determined by legal proceedings or arbitration	This makes adjudication binding until legal action or arbitration – at the end of the contract
108 (3A)	Adjudicator can correct his decision	Amendment from the 2009 Act. A new "ninth" point

108 (4)	**The contract shall provide in writing that it protects adjudicator from action by a party**	NB: This is not Statutory protection, so does not protect the adjudicator from action by third parties, for example, in tort. Adjudicators should carry professional indemnity insurance to provide for this possibility
108 (5)	**Scheme for construction contracts to apply if the contract does not comply with (1)–(4) above**	The scheme (a Statutory Instrument) provides standard clauses allowing for the above points. SBC/Q2016, MWD2016 and DB2016 all invoke the Scheme
108A	**Adjudication costs: effectiveness of provision**	Amendment from the 2009 Act
108A (1)	**Applies to contractual provisions made between the parties concerning the allocation of costs of adjudication**	For example, a common scam requiring payee to pick up costs of adjudication
108A (2)	**The contractual provisions referred to in subsection (1) are ineffective unless:**	Outlaws such scams!
(a)	1 Made in writing 2 Contained in the construction contract, and 3 Confers power on the adjudicator to allocate fees between the parties	For example then, the clause prohibits a requirement for another party to pick up the costs of the adjudication unless it also allows the adjudicator to allocate the costs!
(b)	1 **Made in writing after the giving of notice of intention to refer the dispute to adjudication**	That is, the parties agree after referral to allocate costs

Payment

109 (1)	Stage payments must be written into a construction contract unless it is less than 45 days in duration (in contract as 45 days or agreed by the parties to be so)	For very small contracts, there is no point in stage payments
109 (2)	May agree amounts and intervals etc.	The Act does not, itself, set down time intervals or dictate directly what is to be paid in a stage payment. This means that there is no obligation to have stage payments at all – the interval for stage payments may be set to the anticipated contract duration or longer!
109 (3)	The relevant provisions of the "Scheme for construction contracts" are to apply if the above provisions are not included in the contract	The Scheme (mentioned above) is more prescriptive and sets intervals (if none stated in the contract) of 28 days for payment and the "value of work performed" for the amounts. SBC/Q2016, DB2016 and MWD2016 all set out in detail payment provisions complying with the Act, so the Scheme does not apply
110 (1)	a There must be in a construction contract a mechanism for calculating payments due at each stage and a date at which they are due – the "due date" b There must also be a final date for payment at which the amounts due must be paid Gap between (a) and (b) subject to agreement by parties	The Act does not lay down what the mechanism must be, so payments may be the traditional value-based amounts or true stage payments. The due dates may be set by reference to stages, so can be flexible as envisaged with true stage payments. Again, the Scheme is more prescriptive, with the due date being 7 days after the 28 day "relevant period" ends (or the work has been done if true stage

		payments) and the final date for payment being 17 days thereafter
110 (1A)	Inclusion of "Pay when certified (on another contract)" provisions prohibited	A 2009 Act amendment, this is most relevant to sub-contracts, where "pay when paid" prohibition in section 113(1) below was circumvented by including pay when certified clauses in contracts. Typically, the certification would be by an architect on a main contract and effectively link payment on a sub-contract to payment on the main contract
110 (1D)	Inclusion of payment timing provisions linked to notice of amount of payment due prohibited	Another 2009 Act amendment. The abuse this stops involves including a contract clause allowing the adjustment of the timing of payments to fit the issue of a notice of how much is due, that is no notice, no interim payment!
110A	Payment notices: contractual requirements	
	1 (a) Payer to give notice to the payee not later than five days after the payment due date, or (b) payee to give notice not later than five days after the payment due date	NB: "Notice" not payment!
	2 Notice specifies: (a) and (b) i The sum that the payer etc. considers to be due at the payment due date ii The basis on which that sum is calculated	This notice sits between the due date and final date for payment

	3 Notice specifies: a The sum that the payee considers to be due at the payment due date b The basis on which that sum is calculated	This provision allows for "self-certification" by the payee, where this provision is written into the contract
110B	Payment notices: Payee's notice in default of payer's notice	
	1 (a) and (b) where the contract requires the payer to give notice but notice is not given 2 the payee may give to the payer notice 3 final date for payment postponed by the number of days late the notice was given	This is "self-certification" by the payee, but in situations where the contract intends the payer to certify
111	Requirement to pay notified sum	
	(1) The payer must pay the notified sum on or before the final date for payment	
	(3) The payer may give payee a pay less notice	That is may make a "set off", provided notice is given
	(4) A notice under subsection (3) must specify: a the sum that the payer considers to be due and b the basis on which that sum is calculated	The "set off" must be quantified on some form of basis. This follows the common law rules on set off in construction contracts
	(5) A notice under subsection (3): a must be given not later than a "prescribed period" before the final date for payment	There must be notice of an impending "set off" – the "prescribed period"

	(7) "prescribed period" means: a such period as the parties may agree, or b the period provided by the Scheme for Construction Contracts	Logically, this period cannot be so long as to bring the notice in front of the due date!
	(9) Where adjudication of the amount is involved, the decision of the adjudicator shall be construed as requiring payment of the additional amount not later than: a seven days from the date of the decision, or b the date which apart from the notice would have been the final date for payment, whichever is the later	
	(10) Payment not required where: a the contract provides for no payment in the event of insolvency, and b the payee has become insolvent after the issue of the pay less notice	
112(1)	Right to suspend performance	
112(2)	Seven days' notice required	
112(3)	Right to suspend ends on payment	
112(4)	Extension of time granted	

113(1)	Pay-when-paid prohibited except when "third party" (e.g. The client) is insolvent	With the 2009 Act amendment, pay when certified is also prohibited
114	The Scheme for Construction Contracts	Allows for making this Statutory Instrument

NOTES

1. 1996 Chapter 53.
2. 2009 Chapter 20.

14

Building information modelling and the JCT contracts

INTRODUCTION

In concept, building information modelling is simple. Gu and London's (2010) definition quoted below is fairly easy to understand:

> Building Information Modelling (BIM) is an IT enabled approach that involves applying and maintaining an integral digital representation of all building information for different phases of the project lifecycle in the form of a data repository. (Gu and London 2010, 988)

Another easily understandable definition, drawing on practice in the USA comes from Azhar (2016):

> With BIM technology, an accurate virtual model of a building, known as a building information model, is digitally constructed. (Azhar 2016, 241)

In its implementation, however, BIM is rather more complex. It is this complexity that has most impact on contract administration using the three JCT forms of contract included in this book. Before illustrating how both the power of BIM and its complexity affects contract administration, it is worth pointing out that modelling is nothing new. People have conceptualised in three dimensions from the beginnings of time and designers are trained to think three dimensionally. Quantity surveyors are also trained to translate three-dimensional figures into costs and construction managers to translate those same figures into a programmed sequence. The difference with BIM is that much of the creative thinking process is encompassed within a single digital representation. This means both that it will no longer be necessary to imagine the figure to be worked on as it will always be available in digital form, but also that many tasks (such as taking off quantities) will be removed, or greatly de-skilled. Additionally, the parametric features of building information models allow interdependencies of elements and sub-models to be organically and automatically integrated when changes are made at any stage of the construction cycle (Barnes and Davies 2014).

Complexity is introduced first by the rather mechanistic division of modelling into levels. Again, Gu and London (2010), drawing on Australian guidance, provide succinct explanations of four levels from zero to level 3. Level zero involves two and three-dimensional design without embracing "object-oriented modelling". Level 1 involves object-oriented modelling, but within one discipline. Level 2 involves object-oriented modelling shared between several disciplines and level 3 involves full integration. "Object-oriented modelling" is at the heart of BIM and is probably best thought of as three-dimensional representation of the building suspended in space, but with the facility to zoom in to very detailed levels of resolution. At present, most architectural, engineering and construction efforts are limited to encouraging level 2 BIM and it is this level which is assumed for the remainder of this chapter.

The development of BIM in the United Kingdom has involved the publication of a number of documents including PAS 1192-2:2013 from the British Standards Institute, the Construction Industry Council protocol for building information modelling (CIC BIM Protocol, Second Edition 2018) and the Architectural, Engineering and Construction Committee BIM Protocol (AEC[UK] BIM Protocol v2: 2012) and BIM Technology Protocol (AEC[UK] BIM Technology Protocol v2.1.1 2015). The protocols define in detail the role and duties of key members of the building team, where BIM is to be used and it is envisaged that one or other protocol will be incorporated in contract requirements.

The JCT has contributed to the integration of BIM into contracts in two practice notes covering collaborative and integrated team working and detailed guidance on incorporating a BIM protocol into the design and build form DB2016. The latter has been brought forward to the contract forms.

AN EXAMPLE USING THE CIC BIM PROTOCOL 2018

The CIC BIM Protocol 2018 has received considerable support in the UK construction industry, particularly in relation to larger organisations. It is a short document consisting of 5 pages of guidance and 10 pages of protocol. Its appendices allow parties to a contract to insert particular arrangements for their contract, thus making the Protocol a working document suitable for incorporation in the main contract "Agreement" (e.g. SBC/Q2016 or DB2016). Should the parties to a building contract using one of the three JCT contracts covered in this book, wish to adopt this protocol, the implications are likely to be as illustrated below. This analysis is based on the headings from the CIC protocol but does not repeat the detailed wording. It should be read in conjunction with the full protocol, available from the CIC.

Protocol requirement	SBC/Q2016	DB2016	MWD2016
Definitions			
Agreement The protocol is to be attached to the "Agreement", the latter being the primary contract to be used. This could be a professional services agreement for a consultant, but for construction is the construction contract	SB/Q2016 allows for naming a protocol at 1.1 and gives precedence to the protocol for the form of Contract Documents in 1.4.6 and 2.8.2	DB2016 allows for naming a protocol at 1.1 and gives precedence to the protocol for the form of Contract Documents in 1.4.6 and 2.7.2	There are no provisions for including a protocol in the form and, if required, it would need to be written in, including giving it precedence over the form where necessary
Built Asset Security Manager It is this individual's responsibility to ensure security of information from start to finish of the project. The Employer is required to make and maintain this appointment, but for smaller projects, it is likely to be filled by an existing member of the team	In SBC/Q2016, this role is likely to be filled by the architect	In DB2016, this role is likely to be filled by the Employer's Agent (if there is one) or the contractor	In MWD2016, this role is likely to be filled by the architect
Employer	The Employer for SBC/Q2016	The Employer for DB2016	The Employer for MWD2016
Employer's Information manager The protocol requires the appointment of an Employer's information manager to control information in the model.	There is no need to name the information manager separately in the	As SBC/Q2016. For smaller projects, this role is likely to be filled by the	As SBC/Q2016 provided, the protocol is identified in the contract as

The role is outlined in a separate CIC "Outline Scope of Services for the role of Information Management" (CIC/INFMAN/S, 2013)	contract if the protocol is included in the Contract Particulars. For smaller projects, this role is likely to be filled by the architect	employer's agent or contractor	an amendment
Employer's Information requirements The protocol allows the Employer to specify the information requirements for the project at Appendix 2.1. This relates to the technical production of the model and is unlikely to be directly of interest to most Employers Conflicts are possible between some aspects of the information requirements and other clauses in JCT contracts (e.g. related to the timing for providing information). The protocol wording takes precedence over contract wording for many BIM matters, so it is unnecessary to amend the latter. However, providing conflicting information will cause confusion and the contracts should be expressly brought into line with the protocol	For a typical SBC/Q2016 project, the architect is likely to drive the information requirements as it is this consultant who has most input to the design. In the case of an established "professional client", requirements may be specified directly by the Employer The Design Submission Procedure identified in clause 1.1 of the form gives precedence to the protocol, when incorporated	For full design and build projects, the information requirements are likely to be driven by the contractor. However, where consultants develop scope designs, they may specify information requirements at the pre-contract stage The Design Submission Procedure identified in clause 1.1 of the form gives precedence to the protocol, when incorporated	As nearly all design is likely to be executed by the architect, the requirements will be related to systems used by that consultant There is no formal design submission procedure for providing information (e.g. for CDP work) in this form, but incorporation of a protocol would allow this to be set out. It is suggested in guidance notes that protocol requirements are included in Employer's Requirements for CDP work

Federated information model			
The protocol envisages that the building information model will have contributions from several designers and constructors, which need to be integrated into a federated model. Depending on the form of contract, the complexity of the federation will vary	Most of the models are likely to be produced (pre-contract) by a consultant architect, structural engineer and, possibly, services engineer. Some models (post-contract) are also possible for contractor designed portions of work	Most of the models are likely to be produced (post-contract) by the contractor and subcontractors, with any consultant architects/ engineers producing simpler scope models (pre-contract) prior to tender	The models are likely to be primarily provided by a consultant architect with small contributions from a structural engineer (pre-contract) and (post-contract) from services subcontractors etc. designing CDP elements of the work
Information particulars The protocol provides a template at Appendix 2 for information requirements including Employer's information requirements, BIM execution plan and project procedures	Incorporated in the contract as part of the protocol	Incorporated in the contract as part of the protocol	Can be incorporated in the contract by amendment as part of the protocol
Levels of definition The protocol envisages that models will be worked up progressively, with responsibility for providing increasing levels of detail passing from one to another team member as the design is developed. Levels of definition are divided	As most of the model is likely to be produced (pre-contract) by a consultant architect, the level of definition requirement is likely to be simple and	For simple design and build projects, the levels are likely to be much the same as for SBC/Q2016, but with the contractor's architect producing most design. For	As with contracts using SBC/Q2016, the consultant architect is likely to have direct control over all levels of the model, with small

into two parts, level of model detail for graphical information and level of information for non-graphical content Provisions for specifying levels of definition at various stages are allowed for in Appendix 1, where a Responsibility Matrix can be appended or separately referenced	within one organisation. Some scope design may be provided by the architect and other consultants for contractor designed portions of work and levels of definition will need to be specified for the scope and detail elements	complex projects, subcontractors may have a larger design input, involving more complex levels. For "scope" design and build projects, where a consultant architect is appointed, there may be a further level prior to tender. The "responsibility matrix" (see below) specifies what level of detail provided at each stage	contributions by a structural engineer and some subcontractors for CDP elements Level of definition requirements will, therefore, be simple
Material Material is information in electronic format. It replaces terminology such as drawings, specification, schedules of work etc.	The contract allows at 2.8.2 the protocol to specify alternative forms of communication, from the traditional	The contract allows at 2.7.2 the protocol to specify alternative forms of communication, from the traditional	Alternative forms of communication should also be expressly specified in the form, perhaps by referring to the protocol
Other project team member It is envisaged that other contracting parties will have similar protocol agreements and will be required to work cooperatively to produce models	Relevant other project team members would be (for a contractor), the architect, structural engineer and	Relevant other project team members would be (for a contractor), any Employer's Agent, consultant architect,	Relevant other project team members would be (for a contractor), the architect, structural engineer and

	other consultants, plus subcontractors, particularly those involved in contractor designed portions of the work	structural engineer etc. producing scope designs plus subcontractors, particularly those involved in design	subcontractors designing CDP elements such as services
Permitted purpose For team members to contribute to models, they need access and permission from those who might hold copyright over their work. The protocol allows this for a "permitted purpose" – a purpose which is consistent with the use of the information. For example, a subcontractor working up a design for pricing or construction purposes would have access to fill in details, but not change scope information	Access to the model is likely to be to consultants, contractor and subcontractors carrying out design. Some access may be required for construction programming and costing purposes for the quantity surveyor and other subcontractors	Access to the model is likely to be much as with SBC/Q2016, but with greater design access, where the contractor and subcontractors are carrying out substantial design	Access to the model is likely to be much as with SBC/Q2016
Project	The project for which SBC/Q2016 is being used	The project for which DBC2016 is being used	The project for which MWD2016 is being used
Project agreement The protocol envisages a whole series of agreements between the Employer and team members, of which the construction contract is	In SBC/Q2016, the agreements are likely to be limited to consultants and the contractor. If collateral	For full design and build projects, the agreement may be limited to the contractor only. Any consultants	In MWD2016, the agreements are likely to be limited to the architect, any structural engineer and

only one. Conceivably, collateral warranties with subcontractors could invoke the protocol, giving the Employer direct control over information requirements for their work	warranties are used, these could also become Project Agreements	carrying out scope design may have separate agreements and collateral warranties may also be Project Agreements	the contractor
Project team member	In SBC/Q2016, the Contractor	In DB2016, the Contractor	In MWD2016, the Contractor
Protocol	Incorporated by naming in the Contract Particulars of SBC/Q2016	Incorporated by naming in the Contract Particulars of DB2016	Incorporated by amending the form
Responsibility matrix As mentioned above (level of definition), this instrument specifies what will be contributed to a model, by whom and when. The matrix is intended to be inserted at Appendix 1 of the protocol	Most of the information in the matrix is likely to be the responsibility of the architect, with support from other consultants. For contractor designed portions, responsibility may change as detail develops and this should be included in the matrix	Most of the information in the matrix is likely to be the responsibility of and produced by the contractor, who may pass responsibility for detail along the supply chain. For "scope design and build" projects, consultants may drive the format and requirements of the matrix	The matrix is likely to be very simple, with few parties involved other than the architect
Specified information	The information required of the Contractor by the	The information required of the Contractor by the	The information required of the Contractor by the

	Responsibility Matrix and in Appendix 2 of the protocol	Responsibility Matrix and in Appendix 2 of the protocol	Responsibility Matrix and in Appendix 2 of the protocol
Detail of the protocol			
Priority of contract documents Makes clear that, in any conflict between the terms of the protocol and the agreement, the agreement prevails. However, if the agreement does not have provision for resolving the conflict, or for several BIM-related matters concerning the obligations of Employer or Project Team Member, then the protocol is stated to take precedence	Clause 1.3 of SBC/Q2016 gives precedence to the Conditions of Contract. Where a protocol is used, terms are modified to allow the provisions of the protocol to determine documentary requirements (e.g. in clause 1.4.6)	As SBC/Q2016	Any modified requirements, for example as to the form of Contract Documents for a project using BIM, would need to be written into the contract as Clause 1.2 gives priority to the Conditions of Contract
Obligations of the employer The protocol requires the Employer to, amongst other things, incorporate the protocol in the agreement, comply with the information requirements, ensure that the information and built asset security manager's roles remains filled and that project team members can share in the common data environment	In SBC/Q2016, this requirement is likely to be delegated to the architect	In DB2016, this requirement is likely to be delegated to the Employer's Agent (if there is one) or the contractor	In MWD2016, this requirement is likely to be delegated to the architect

Obligations of the project team member In short, the protocol requires the "project team member" (in building contracts, the contractor) to provide building information modelling in accordance with the protocol. Specifically to: • Produce • Share and • Use, information • Arrange for incorporation of the protocol in subcontracts • Comply with project security requirements	The contractor is tied into BIM by this obligation and is required to also tie in subcontractors	The contractor is tied into BIM by this obligation and is required to also tie in subcontractors	The contractor is tied into BIM by this obligation and is required to also tie in subcontractors
Electronic data exchange The protocol puts the compatibility of electronic data delivered in accordance with the protocol at the risk of the Employer. Also, once data have been transmitted, the Project Team Member is not liable for any corruption, unintended amendment or alteration	In SBC/Q2016, this protocol requirement puts the onus on the primary consultant (the architect), to make sure that contractor supplied input to models is compatible and does not become corrupted	In DB2016, responsibility for ensuring data integrity is likely to pass respectively to an Employer's Agent, a consultant "Scope design" architect or to be retained, effectively, by the Contractor	As SBC/Q2016

Use of information This part of the protocol regulates copyright and similar agreements. It requires the Project Team Member (the Contractor) to give necessary access to develop material but restricts changes unless specifically authorised by the Employer's Information Requirements. Material may be used by Other Project Team Members for Permitted Purposes, but not otherwise. For detail on the provisions, see the protocol itself	As the protocol is incorporated in SBC/Q2016 by reference there is no need to amend the form	As SBC/Q2016	As SBC/Q2016
Liability in respect of proprietary material The protocol limits the liability of the Project Team Member and Employer for modifications to material other than for copying and using the material for a Permitted Purpose This relates to the model only and not to the substance of any design. A subcontractor, therefore, providing an inadequate design, would still be liable if it failed	Incorporated in SBC/Q2016 by reference to the protocol	Incorporated in DB2016 by reference to the protocol	Incorporated in MWD2016 by making reference to the protocol in an amendment

Remedies – security The protocol gives the Employer power to take action in the event of an actual or potential breach of security related to project information and, in particular information defined as sensitive. Action can include, in extreme cases, termination without notice	Incorporated in SBC/Q2016 by reference to the protocol	Incorporated in DB2016 by reference to the protocol	Incorporated in MWD2016 by making reference to the protocol in an amendment
Termination This part of the protocol states that the protocol shall continue to apply following termination of the Project Team Member's employment under the Agreement. This provision is consistent with the JCT contracts, which leave the contract in place, but terminate employment in the event of major breach by or insolvency of the Contractor	Incorporated in SBC/Q2016 by reference to the protocol. Special provisions may be needed to ensure that material remains available after termination (particularly related to insolvency)	As SBC/Q2016	As SBC/Q2016
Appendix 1			
Responsibility matrix This is a blank page intended to be populated with relevant material or by inserting a reference	For SBC/Q2016, this is likely to be produced by, or on behalf of the consultant architect	For DB2016, this is likely to be produced by the Contractor (or possibly Employer's Agent or consultant architect, if scope design and build is to be used)	For MWD2016, this is likely to be produced by the consultant architect

Appendix 2			
Information particulars This is a detailed pro-forma, with insertions intended to cover the Employer's Information Requirements, BIM Execution Plan and detailed project procedures	In SBC/Q2016, the architect is likely to be the relevant authority producing this. For smaller projects, the architect may also be the information manager	In DB2016, the Contractor is likely to be the relevant authority producing this. Where an Employer's Agent or scope architect is used, this individual may compile the information (or require that a separate information manager does so)	In MWD2016, the architect is likely to be the relevant authority producing this. He/she is also likely to be required to take on the role of Employer's information manager
Appendix 3			
Security requirements This is also a detailed pro-forma, with insertions intended to cover what information is sensitive and how it is to be handled, project-specific security requirements and Employer's policies and procedures. The name of the Built Asset Security Manager is to be inserted in this section	In SBC/Q2016, the architect is likely to be responsible for assisting in identifying these requirements and, for smaller projects, also taking the role of the Built Asset Security Manager	In DB2016, the Employer's Agent or Contractor is likely to be responsible for assisting in identifying these requirements and, possibly, also taking the role of the Built Asset Security Manager	In MWD2016, the architect is likely to be responsible for assisting in identifying these requirements and taking the role of the Built Asset Security Manager

PRACTICAL CONSIDERATIONS

Building information modelling is often portrayed as a cooperative method of producing building information involving input from several parties. This is reflected in the relative complexity of Responsibility Matrices hitherto produced. Cooperative and incremental development of information does not appear to sit well with the JCT forms of contract considered in this book. These contracts envisage a clear division of responsibility with the Employer setting out requirements to which the Contractor responds. In the traditional forms of contract SBC/Q2016 and MWD2016, it is intended that the Employer alone, through a consultant architect, will produce all the design for the building. The contractor's input is to coordinate resources and build to the design. This arrangement may be complicated where sections of an otherwise architect designed project are to be designed by the contractor (or subcontractors) as "contractor designed portions". Design stage cooperation may be required for this arrangement and needs to be allowed for in any modelling. With the design and build form of contract, DB2016, it is easier to promote cooperation once the contractor is appointed, as the latter has control over both design and construction. However, where consultant scope designers are used, it is necessary to integrate scope and detail and this may involve conflicts.

In practice, these problems can be fairly easily tackled provided that the key individual driver (or champion) of the BIM process is confident with the modelling process itself. Taking each of the forms in turn, the requirements of BIM can be met as follows:

Traditional contracts using SBC/Q2016

This form envisages full design by an architect, supported by other specialist consultants including structural and services engineers and quantity surveyors. Provided the form is used as traditionally intended (all design remains the responsibility of consultants), the modelling process is largely a pre-tender activity driven by the architect. The architect must be sufficiently proficient to be able to act as Employer's Information Manager, or to be capable of authorising a separate individual. In cooperation with the quantity surveyor, the architect can ensure that a suitable protocol is incorporated in the contract form and mentioned in tender documents. The contractor will, therefore, have been put on notice that it will be required to supply relevant information to the model once appointed. The demands of modelling adequately mean that information will need to be sorted out before tender and there will be less scope for design development or contractor contributions. This is not necessarily problematic as carefully prepared and complete designs are usually a spur to speedy and efficient construction.

If contractor designed portions are to be used, it will be necessary for the consultant architect or other consultant to have scoped out the element prior to tender as, perhaps, a level 1 model. The appointed contractor, or relevant subcontractor, will then be required to contribute design at a later level. Tendering and costing can be dealt with in the model in the traditional process of providing provisional sums or quantities. If it is necessary to incorporate specialist subcontractor design in a model pre-contract (e.g. in order to contribute to architectural or engineering design), then some form of nomination and prior appointment o the specialist will be required. This is allowed for in provisions to name specialists in SBC/Q2016 and prior naming of subcontractors in DB2016[1].

Traditional contracts using MWD2016

The process for minor works is likely to be very similar to larger works using SBC/Q2016. Considerable demands may be put on architects specialising in small works, as there is unlikely to be sufficient fee to appoint a separate Employer's Information or Built Asset Security Manager (unless this can be arranged on a part-time or bureau basis). The architect will need to be sufficiently proficient at modelling to be able to personally construct the model and pass to other project team members. Other than this demand, the process is much the same as with SBC/Q2016. There is a certain element of variability inherent in small works, where documentation is less clearly defined. However, BIM should simplify this and may enable the easy production of uniform comprehensive information at the level of full bills of quantities.

Design and build contracts using DB2016

Pure design and build contracts, where the scope and detailed design are produced by the Contractor, are ideal for cooperative development of models. Provided that sufficient information can be produced to agree a price with the Employer, much of the scope and detail can be worked up cooperatively, as envisaged in the CIC protocol. For most projects, it is likely that the Contractor will employ a specialist Information Manager and may develop considerable expertise in model development and use.

Where scope designs are produced by consultants and/or Employer's Agents are used, the detailed requirements for BIM may be specified before the appointment of the contractor. Some consultants retain supervisory duties and this will need to be allowed for in the Responsibility Matrix and will complicate development. However, overall, modelling should be both simpler and more integrated that for traditional contracts.

TERMINOLOGY

There is a time-honoured terminology associated with building projects, related to conceptual outputs. This terminology includes:

- Drawings
- Bills of quantities
- Schedules of rates
- Specifications
- Schedules of works

- Employer's requirements
- Contractor's proposals
- Contract sum analysis

In building information modelling, these outputs are replaced with level 2 models (possibly also including cost and schedule models). However, there is little need to change traditional terminology in day-to-day use for two reasons:

1. Outputs from models can easily be configured to fit the former hard copy documentation. In other words, a cost data printout at (for example) stage 2 of model delivery can be in the form of bills of quantities.
2. The protocol terminology is included in the contract by reference to the protocol. Where information is to be provided in a form other than specified in the contract, provided it is clear in the details of the protocol, then this provision will comply with the contract.

NOTE

1. See Chapter 9 for more details of these subcontracting arrangements.

BIBLIOGRAPHY

Architectural, Engineering and Construction Committee (2012) *AEC(UK) BIM Protocol*, AEC, London, UK

Architectural, Engineering and Construction Committee (2015) *AEC(UK) BIM Technology Protocol*, AEC, London, UK

Azhar S (2016) Building Information Modeling (BIM): Trends, Benefits, Risks, and Challenges for the AEC Industry, *Leadership and Management in Engineering*, July, 241–252

Barnes P and Davies N (2014) *BIM in Principle and in Practice*, ICE Publishing, London, UK

British Standards Institution (2013) *PAS 1192-2:2013 Specification for information management for the capital/delivery phase of construction projects using building information modelling*, BSI Standards Limited, London, UK

Construction Industry Council (2018) *Building Information Model Protocol*, 2nd edition, Construction Industry Council, London, UK

Construction Industry Council (2013) *Outline Scope of Services for the Role of Information Management (CIC/INFMAN/S, 2013)*, Construction Industry Council, London, UK

Gu N, London K (2010) Understanding and facilitating BIM adoption in the AEC industry, *Automation in Construction*, 19, 988–999

JCT (2016) *Building Information Modelling (BIM) Collaborative and Integrated Team Working Practice Note*, Sweet and Maxwell, London, UK

JCT (2019) *BIM and JCT Contracts Practice Note*, Sweet and Maxwell, London, UK

Glossary

Advance payment bond

An on-demand bond provided by the Contractor in favour of the Employer in consideration for advance payment of an amount. The amount is normally for an element of a building, such as a prefabricated module, and the amount of the bond is reduced as the advance is repaid on construction.

All-risks insurance

Insurance covering all risks with exceptions. Unless an excluded risk is mentioned in the policy, it is covered by the insurance.

Approximate quantities

Measured items of work where either the exact nature of the items or their exact quantity is not known. Once these are determined (usually post-contract), the exact details are substituted and costs are adjusted.

Approximate estimate

An estimate of the cost of building work, made by a consultant (for example a QS or Architect) or builder as a guide and not intended to form a legal offer. Sometimes, where builders have control over design, they will submit tenders based on approximate estimates.

Architect

The lead consultant in traditional contracts based on SBC/Q2016 and MWD2016. The Architect has both agency duties on behalf of the Employer and impartial certification duties for key aspects of construction (for example certifying practical completion).

Averaging
Reducing the amount of a claim by an insurance company pro-rata to the extent that the work or building is under-insured. Insurance premiums for buildings and works are always set on a full-value basis (rebuilding cost, or contract sum plus fees) irrespective of the loss incurred.

Back-to-back contracts
Contracts in a supply chain that have substantially the same terms. For example, SBCSub/C2016, the JCT standard subcontract form has similar terms to SBC/Q2016, the traditional main contract form.

Bill rates
Rates for items of work inserted in the contract Bills of Quantities.

Bills of quantities
A Contract Document with SBC/Q2016 containing all cost significant matters in written form. These include preliminary items, detailed and general specifications (called preambles), measured items of work and allowances for unknown work (provisional sums).

Bills of reductions
Adjustment bills of quantities produced by the QS (SBC/Q2016) with the intention of reducing the level of a tender. It contains items and quantities to be omitted or reduced in the tender and cheaper items to be added as substitutes. The financial effect of the reductions is calculated, deducted from the tender and put forward by the Employer as a counter-offer.

Bond
A financial guarantee put forward by a surety company or bank and payable on the occurrence of an event, such as valid termination of a contractor's employment under a contract (performance bond), or failure to repay an advance payment by a contractor to an employer (advance payment bond). The bond is taken out by the contractor, priced into a contract and paid by the Employer. See also Chapter 10.

Buildability
The extent to which a building or element is easy and efficient to build.

Builders quantities
Measured items for pricing produced internally by builders (rather than by a consultant QS). These quantities may not be produced in strict accordance with the requirements of NRM2.

CDM Regulations
UK Government (2015) The Construction (Design and Management) Regulations 2015, SI 2015 No 51, The Stationery Office, London. This governs several health and safety matters on construction projects.

Certificate
A document produced by the Architect impartially indicating the status of a project or elements of a project (relates to traditional SBC/Q2016 and MWD2016 contracts). For example, an interim payment certificate indicates the value of work properly executed in a given time period.

Clerk of works
A site inspector engaged by the Employer, under the supervision of the Architect (relates to traditional SBC/Q2016 and MWD2016 contracts).

Collateral warranty
A contract running alongside others in a supply chain used in order to create a direct liability between two parties. For example, between an occupying tenant and a subcontractor – in exchange for a nominal payment, the providing party (subcontractor) warrants performance of its work to the benefiting party (tenant).

Concurrent delay
A cause of delay occurring alongside another cause. For example, one cause might be exceptionally adverse weather and another might be inability to obtain building materials on time.

Concurrent expense
A cause of additional expense occurring alongside another cause. For example, the cost of uneconomic working of labour might be caused either by a variation or poor organisation.

Construction Act
See Chapter 14. Housing Grants, Construction and Regeneration Act 1996 – Part 2 as amended by the Local Democracy, Economic Development and Construction Act 2009.

Contract Administrator
The alternative term for Architect in SBC/Q2016 and MWD2016, where the individual is not an architect registered under the Architects Act 1997.

Contract documents
The documents evidencing details of the contract.

Contract particulars

The variable elements of the Conditions of Contract, specific to the project. In SBC/Q2016, DB2016 and MWD2016, these particulars are collected in a section towards the beginning of the Conditions of Contract.

Contract rates

Rates for items of work inserted in the contract documents. For SBC/Q2016, these are Bill Rates, but similar rates may be used in Schedules of Rates (MWD2016), or a Contract Sum Analysis (DB2016).

Contract-sum analysis

The pricing document associated with DB2016. May range from a simple breakdown of costs at stages of construction, through schedules of rates, to full bills of quantities.

Contractor designed portion (CDP)

In SBC/Q2016 and MWD2016, an element of a building designed and constructed by the Contractor and dealt with in a separate section.

Contractor's proposals

The offer from a tendering contractor in DB2016. May include a full design and specification with associated costs in response to a written or verbal brief or include detailed design, specification and costs in response to an outline design.

Cost plan

A financial plan of the approximate distribution of costs of a building project. Associated with consultant produced approximate estimates.

Cover pricing

A tender submitted at a high level which is not expected to be competitive, but which will show interest from the tenderer and/or produce abnormal profits if successful.

Cube test

A test of batch concrete consisting of a small sample cast into a cube, cured and then crushed to record its strength.

Dayworks

Payment on the basis of the agreed direct cost of labour, materials and plant. Labour is paid at an all-in hourly rate, materials and plant at invoice prices. An agreed percentage is added to direct costs to cover general overheads and profit.

Defined work

Associated with provisional sums. Work which fits the definition given in NRM2 for "defined work" (see Chapter 4 page 64) – sufficient details of the work are provided for a tendering contractor to estimate and include in a tender the cost of associated preliminary items.

Discounted cash flow (DCF)

The present value of a flow of funds or costs in the future, assuming financing has an interest cost.

Disputes advisor

A contract appointment of an individual to assist in settling project disputes as and when they arise.

Disputes board

A committee of disputes advisors acting to settle project disputes.

Due date

The date at which interim or final payments are due to be paid. Terminology is from the Construction Act.

Due diligence

A thorough financial examination of a proposal. In tendering situations involves an employer or a consultant inspecting submitted tender documents to ensure that they have been properly completed and a contractor's offer is soundly based.

Employer's agent

In DB2016, an individual acting as the Employer. The agent will normally be drawn from one of the construction professions as the technical requirements on the Employer in DB2016 require considerable specialist expertise.

Employer's requirements

In DB2016, the briefing provided by the Employer for the Contractor to expand on and price. May range from a simple verbal or written performance requirement to full drawings and specification.

Estimated prime cost

An approximate estimate of the cost of building work. Used in cost plus contracts as a basis for calculating the cost of overhead additions to prime costs and as a basis for setting target costs.

Expert (or professional) client
A client who builds regularly and is familiar with the managerial processes involved.

Factor costs
The costs of labour, materials, plant, general overheads and profit.

Fair valuation rates
Price rates for varied or extra work in a final account based on an objective valuation, usually agreed by negotiation between parties. Normally used where contract or pro-rata rates are not available.

Final date for payment
The date at which a contractor will be paid. This is normally later than the Due Date (see above) to allow the Employer time to give notice of an impending set-off under the terms of the contract. Terminology is from the Construction Act.

Framework agreements
An agreement to let a series of individual projects to a single contractor.

Front loading
Increasing price rates for carrying out work early in a construction sequence and decreasing price rates for carrying out work late in a construction sequence, without increasing an overall tender figure. Done, where interim payments are made, to improve contractor cash flow.

FRI lease
A building lease, where the lessee is required to fully repair and insure the building. Accordingly, the lessee cannot ask the lessor to make repairs.

General overhead
Cost items of a general nature, which cannot be priced to an individual project. Examples are head-office administrative expenses.

Implied term
A contract term not written down in the contract documents, but based on general law principles.

Key performance indicator
An indicator of performance of a contract, often not based directly on financial indicators, but on more qualitative criteria.

Key rates
Price rates for significant items making up a considerable proportion of the total cost of building work.

Latent defects
Defects of construction not apparent on completion of work.

MEAT
Most economically advantageous tender, taking into account costs in use and intangible benefits evaluated in economic terms.

Mutual dealing
A process of working out the net indebtedness of a firm to another by allowing for all debts and credits between the firms.

Named risks insurance
Insurance covering risks named in the policy. If the risk is not named, then it is not covered. In JCT contracts named risks are referred to as "Specified Perils".

Named specialist
A specialist subcontractor either named in Contract Documents (Pre-named) or named against a provisional sum (Post-named) – relates to SBC/Q2016 Schedule 8 para 9.

Named subcontractor
A subcontractor or subcontractors named in a contract. For details of named subcontractors, see Chapter 8.

Nominated subcontractor
A single subcontractor nominated by an Architect – associated with the JCT98 and earlier forms of contract.

Notional account
The final account for a project, where the contractor has become insolvent during the progress of work, as if that contractor had continued to completion.

Notional cost with EPC
Associated with cost-plus contracts. The actual cost for building work assuming that the Employer had not increased the cost of work. It is compared with the estimated prime cost of the project for agreeing target payments.

Notional quantities

Associated with measured term contracts. Approximate quantities calculated by a consultant QS and multiplied by rates provided by tendering contractors to give a notional tender for carrying out work. Used to compare tenders.

Novation agreement/contract

A new contract with the same content but different parties. Used loosely in two circumstances – the re-engagement of an architect working on a project for a client, to work with the contractor on the same project and the re-engagement of a contractor on a project after it has become insolvent to work with the same client.

Open book

A process of negotiation based on the contractor revealing all details of pricing.

Parity of tendering

The process of treating all tendering contractors equally and fairly.

Partnering

A form of tendering designed to foster cooperative performance. Divided into single project partnering, where contracting parties enter into a cooperative performance agreement in addition to the construction contract, and strategic partnering, where parties establish a cooperative relationship over several projects, in addition to separate construction contracts for each.

Pay less notice

A notice from a paying party to a payee specifying intent to pay less than an amount stated in a payment statement (DB2016) or certificate (SBC/Q2016, MWD2016). The payment statement/certificate is issued on the due date and is due on the final date for payment subject to the prior issue of a pay less notice. Terminology is from the Construction Act.

Performance specification

A specification for a building (or part of a building) based on performance requirements – for example "a school to accommodate 200 students".

Post tender checks

Checks on submitted tender documents and tendering contractors after submission of the tenders.

Preferred bidding

Revealing to a preferred tendering contractor or subcontractor details of pricing of a rival tenderer, with a view to having the price matched or bettered.

Preliminaries

General items of cost significant information relevant to a project as a whole (as opposed to specific measured work and specification items). In SBC/Q2016 contained in a separate Preliminaries section. The contractor will price specific cost items as Site Overheads against the Preliminaries items.

Present value costs

See DCF above – the present value of a flow of funds or costs in the future, assuming financing has an interest cost.

Pre-tender checks

Checks on tendering contractors before submission of the tenders. May include taking references from previous employers and architects and financial checks.

Price ringing

A process where tendering contractors collude to set tender prices, usually selecting the contractor to submit the lowest bid.

Professional client

A client whose business includes placing building contracts – for example a property developer, or Government department.

Pro-rata rates

Price rates based on contract rates where there are no exact equivalents in the contract documents.

Provisional sums

Lump-sum allowances for items of work which have not been described or quantified in detail. The provisional sum is omitted in the final account. Any work ordered against the provisional sum is measured, valued and added in the final account.

Quantity Surveyor

The person named in SBC/Q2016 as the Quantity Surveyor, who has duties of measurement and valuation defined by that form.

Re-rating

Changing the valuation of price rates for varied or extra work in a final account from contract rates to pro-rata or fair valuation rates. Employed where the circumstances for carrying out varied/extra work have changed.

Reservation of title

A term in a sale agreement reserving ownership of goods to the seller until paid by the buyer.

Retention bond

An on-demand bond provided by the Contractor in favour of the Employer in consideration for the latter not holding a financial retention on interim payments. The bond increases in value as work is executed and becomes payable should the Contractor not correct snagging defects.

Reverse auctioning

A form of bidding, where tenderers provide prices at an auction, which decrease in value until no further reduced offers are made. Usually associated with subcontractor tenders.

Schedules of rates

Price rates for carrying out items of work per unit quantity. These may be collected into schedules specifically constructed for a project or may be published Schedules provided for use by public clients. In measured term contracts, schedules are used for tendering purposes. For other contracts (for example using MWD2016), schedules may be used for valuing variations.

Sensitivity analysis

A financial analysis measuring the extent to which a change in one or more factors may affect cost/values elsewhere. For example, an analysis might be conducted to assess the extent to which potential increases in excavation quantities and associated prices might make a contractor's tender less competitive than the same increases with another contractor.

Site overhead items

Cost significant items on a building site, which are not associated with the physical permanent work – for example the cost of scaffolding. In SBC/Q2016, priced by the Contractor against Preliminary items provided in Bills of Quantities.

Slump test

A test of a batch of unset concrete based on the extent to which a sample cone of concrete "slumps", when the cone former is removed.

Snagging defects

Defects apparent towards the end of construction – usually of a minor nature associated with finishing items.

Special project vehicle (SPV)

In construction procurement, a company set up specifically to execute a single construction project. Associated with PFI contracting, a typical SPV would include a contractor, finance company and facilities management company. Each would hold shares in the SPV.

Specified Perils insurance

See named risks above.

Specialists

Firms carrying out sections of building work based on specialist tasks – for example bricklaying, groundworks or steel erection firms.

Statement

A document produced by the Employer indicating the status of a project or elements of a project (relates to DB2016 design and build contracts). For example, a final statement indicates the final account cost of the project. Is the equivalent to "Certificate" in SBC/Q2016.

Statutory adjudication

Adjudication in accordance with the Construction Act – parties to construction contracts have the right to use this process for disputes arising under the contract.

Statutory Undertakers

Specialists carrying out work in pursuance of a Statutory Duty. For example, a gas network company installing the main gas supply to a building. They are not contracted to the Employer or Contractor but work and are paid in accordance with the relevant Statute.

Subrogation clause

A clause in a contract between insured and insurance company allowing the latter to use the name of the former in pursuing a party for an insured loss. For example, where the Contractor negligently destroys building work insured by the Employer, the insurer may use the subrogation clause to sue the Contractor.

Tenant

An individual or company with a limited term interest in land (a lease) granted by a land owner.

Tender

A procedure in selling or offering to carry out work where all offers are submitted at or before a time stated by the buyer.

Tenderer
An entity submitting an offer by tender. For building work, the tenderer is a building contractor.

Tender documents
Documents indicating the intentions of the Employer in a building project.

Tender ratio
For a particular contracting firm, the number of projects for which tenders are submitted divided by the number won.

Undefined work
Associated with provisional sums. Work which fits the definition given in NRM2 for "undefined work" (see Chapter 4 page 64) – sufficient details of the work are not available for a tendering contractor to estimate and include in a tender the cost of associated preliminary items. Should such work be required, an addition will be made for preliminary items.

Valuation rules
In SBC/Q2016 and DB2016, the rules for valuing variations – values are based on contract rates where possible, if not, on pro-rata rates or by fair valuation.

Whole life cost assessment
An assessment of the cost of a building based on construction, running, maintenance and demolition costs. The latter are discounted using a DCF process to Present Values for comparison purposes (for example at tender stage between different bids from contractors).

Working drawings
Drawings sufficiently detailed for construction.

Index

acceleration 82, 85, 90, 96
accepted defects 63
additions (variations) 70, 76, 81, 86, 161
adjudication (in disputes) 64, 74, 83, 97, 139, 173, 174, 180, 184, 186, 188, 191–203, 231
Administrator (in liquidations) 116, 117
advance payment 61, 66, 156, 221, 222
advance payment bond 221, 222
all risks (insurance) 149, 151, 153, 154, 221
approximate estimate 19, 221, 225
approximate quantities 14, 54, 77, 82, 84, 221, 228
arbitration 4, 94, 99, 177, 180–185, 188, 3, 97, 102, 184, 186, 187–193, 197
architect's satisfaction 78, 171
arithmetical check 45
articles of agreement 42, 49, 52
assignment 177
attendance claims 106
averaging (of insurance claim) 154, 222

back-to-back (contract terms) 137, 141, 222
balance of adjustable work 168
bankruptcy 109, 115, 116, 118
base date 164–167
battle of the forms 140, 141

BIM Protocol 4, 48, 52, 206, 220
binding resolution (of disputes) 184, 185, 188
brief (from client) 10, 11, 15, 43, 52, 224
buildability 33, 34, 222
builders quantities 6, 222
building information modelling 4, 51, 205–220
building regulations 7, 79, 172, 178
building standards 25, 39, 79
building surveyor(s) 2, 52, 152, 155
business risk insurance 153

cash flow 1, 43, 44, 46, 47, 226
caveat emptor (in buying buildings) 126
CDP Analysis 48, 76
chain of liability 125, 128, 130, 138
changes (variations) 73, 79, 80, 165
CIMAR (Construction Industry Model Arbitration Rules) 192, 193
CIS (Construction Industry Tax Deduction Scheme) 191
civil engineer 6
claims (contractual) 3, 65, 67, 71, 78, 80, 82, 85, Chapter 6
claims made (basis for insurance) 152
clerk(s) of works 171, 177, 179, 223
Code of Practice for testing work 170, 175, 179

234 *Index*

collateral warranty 128, 129, 138, 223
competent person in charge 170
completion bill (of quantities in liquidations) 119
completion date 77, 82, 88–90, 92–94, 96, 98
conceptual design 11
conciliation 184, 185, 192
concurrent delay 105, 223
concurrent expense 105, 223
conditions of contract 42, 49, 213, 224
consensual resolution (of disputes) 184
consideration 37, 98, 128, 141, 184, 185, 221, 230
Construction Act 64, 66, 67, 71, 78, 92, 139, 140, 150, 180, 181, 186, 188, 191, Chapter 13, 223, 225, 226, 228, 231
CDM (Construction Design and Management Regulations) 52, 223
construction management 7, 17, 22, 33, 35, 119, 196
consultant architect 11, 21, 32, 170, 209, 210, 216
contra proferentum (interpretation of indemnity clauses) 145
contract bills of quantities 48, 49, 50, 222
contract document 48, 78, 222
contract drawings 48, 49, 89
Contract Particulars 42, 55, 62–66, 87–91, 94, 96, 105, 129, 141, 146, 148, 152, 157, 158, 163–167, 191, 192, 208, 212, 224
contract specification 80
contract sum 54, 55, 59, 73, 76, 77, 78, 84, 85, 90, 91, 95, 96, 97, 103, 123, 133, 147, 150, 151, 154, 155, 164, 169, 174, 175, 178, 180, 222, 224
contract sum analysis (CSA) 52, 53, 67, 79, 97, 220
contractor designed portion (CDP) 48, 75, 76, 129, 141, 152, 171, 172, 208, 209, 210, 211, 224
contractor's proposals 48, 52, 53, 75, 79, 80, 179, 180, 220, 224
core (tests) 179
cost plan (cost planning) 15, 45, 224

cost reimbursement 8, 17, 119, 120
cover pricing 26, 224
creditors 59, 115, 116, 121, 122
critical path 105
CSCS (Construction Skills Certification Scheme) 177
cube test(s) 178, 179, 181, 224

date for completion 88, 90, 92–94, 96, 100, 105
date of possession 87, 105
daywork(s) 56, 77, 82, 111, 165, 224
deed (contracts) 50, 79, 127, 152
default (of parties to contract) 111, 112, 114, 133, 137, 141, 147, 150, 152, 156, 157, 162, 164, 192, 201
defect(s) 3, 12, 23, 63, 66, 78, 89, 91, 110, 111, 120, 125–130, 134, 138, 156, 157, 171, 174, 177, 180, 227, 230
defective work 64, 67, 83, 106, 137, 138, 180
deferment 91, 103
deferred (possession) 91
defined work 78, 81, 225
delay(s) 20, 48, 65, 87, 89, 92–94, 98–102, 104–106, 114, 120, 124, 128, 131, 133, 137, 138, 141, 153, 155, 170, 175, 178, 223
delay notice 92
design, manage and construct 20, 21, 35, 60
detailed design 6, 11, 12, 13, 14, 34, 169, 198, 219, 224
discounted cash flow 225, 229, 232
dispute prevention mechanisms 184
disputes adviser 188
disputes board 188
disturbance (of regular progress) 98, 102, 103
domestic subcontractor 133
drawing and specification contracts 9, 12, 13
due date 62, 78, 92, 199–202, 225, 226, 228
due diligence 40, 45, 225

early use 89, 95, 153, 154

early warning notices 187
electronic data exchange (in BIM) 214
elemental cost breakdown 52
elemental format (of final account) 82
employer's final statement 80, 90, 231
employer's requirements 48, 52, 53, 57, 67, 75, 79, 80, 89, 129, 133, 138, 141, 146, 147, 180, 208, 220
end-point financing 60
errors 14, 28, 29, 31, 43–47, 50, 75
estimated prime cost (EPC) 19, 20, 22, 225, 227
estimating 7, 8, 14
ex-gratia claims 98
exclude individuals (from works) 170, 177
existing structure 114, 149
expert determination (of disputes) 184–186
express terms 59, 74, 87, 97, 150
extension contracts 27–29, 35
extension(s) of time 3, 77, 78, 79, 82, 87–94, 96, 99, 102, 104, 105, 163, 170, 175, 202

facilities management 9, 231
factor cost 32, 54, 55, 100, 156, 161, 162, 164, 167, 226
fair assessment 82
fair valuation 76, 102, 226, 229, 232
fast-tracking 1
federated model (in BIM) 4, 209
final adjustment 90, 91
final certificate 74, 78, 80, 83, 90, 92, 127
final account Chapter 4, 45, 54, 61, 90, 120, 123, 124, 147, 226, 227, 229, 231
final date for payment 62, 78, 199–202, 226, 228
final payment 3, 13, 78, 83, 92, 169, 171, 174, 180
final statement 80, 90, 231
financial adjustment 79, 175
financial statement 84
fixed fee 18, 19, 35
fluctuations Chapter 10, 3, 71, 83, 85
form of tender 40

formula fluctuations 162, 167, 168
framework agreement 30, 31, 35
FRI lease (full repairing and insuring lease) 126, 226
front load(ing) 44, 46, 226
frustration 110
fundamental breach (of contract) 109–111

general overhead 21, 99, 106, 226
guaranteed maximum price 21, 35, 60

Hudson's formula 99, 106
hybrid contracts 8

implied term(s) 73, 74, 97, 178, 226
indemnity Chapter 9, specifically 143–146
index, indices for price adjustment 167, 168
information management (in BIM) 57, 208, 220
information requirements (in BIM) 208, 209, 212, 213, 215, 217
insolvency Chapter 7, 3, 140, 143, 147, 155, 180, 202, 216
Insolvency Act 115, 116, 118, 124
instructions 40, 63, 74, 75, 80, 81, 94, 95, 98, 101, 170, 171, 175, 176, 179
insurance Chapter 9, 3,13, 14, 42, 49, 51, 57, 59, 63, 66, 68, 89, 95, 113, 114, 117, 118, 120, 198, 221, 222, 227, 231
interest charges 98, 101
interim application (for payment) 67
interim certificate(s) 14, 62, 64–66, 103, 150, 156
interim payment(s) Chapter 3, 15, 42, 56, 110, 121, 156, 157, 173, 174, 180, 200, 223, 226, 230
interim valuation(s) 61, 62, 64, 67, 84
invitation(s) 24, 25, 37–40, 51,139

joint names (insurance policies) 154, 159

key performance indicators 30, 226
key rates 33, 227
knock-on 106, 77, 138

landscape architects 6
latent defects 3, 12, 23, 125, 128, 129,130, 227
letter of acceptance 42
levels of detail (in BIM) 209, 210
life-cycle costs 51
limitation period 50
limitation of liability (clauses in material supply contracts) 128
Limited Liability Partnerships Act 115
liquidated damages 42, 62, 66, 78, 87, 89, 91, 92, 95, 96, 99, 153, 174, 180
liquidation 59, 98, 116–118, 120–123, 144, 145
litigation 97, 184, 187–190, 192
local authority/local authorities) 7, 9, 148
loss adjustment 155
loss and expense 63, 65, 77, 78, 98, 103–105, 113, 162, 163
loss of profit 47, 98, 100, 101, 102, 114, 115
lump sum 8, 9, 12, 16, 17, 18, 20, 21, 23, 38, 40, 59, 78, 99, 119, 120, 161, 229

making good defects 78, 177
management contracting 8, 20, 22, 35
materials off site 63, 65, 121, 153
materials on site 63, 64, 65, 67, 70
measured term contract(s) 9, 15, 16, 35, 228, 230
measurement and valuation 61, 62, 67, 76, 77, 79, 229
mediation 139, 174, 180, 184–188, 190, 192, 193
most economically advantageous tender (MEAT) 51, 53, 227
mutual dealing (in liquidations) 117, 121, 227

named sub-contractor 133, 138, 141, 227
negligence 3, 79, 115, 131, 145, 146, 151, 154
negotiated contract 24, 28, 30, 32, 33, 35
negotiating claims 104

negotiation 1, 11, 31–34, 51, 54, 55, 56, 83, 103, 119, 183, 184, 226, 228
neutral evaluation 184, 185
New Rules of Measurement (NRM) 38, 56, 81, 167, 178
no negligence (insurance) 146
nominated sub-contractor 133, 137, 227
notional account (in liquidations) 120, 227
notional cost(s) 19, 101, 121, 227
notional quantities (with measured term contracts) 16, 228
novated sub-contractor 134
novation (in liquidations) 118, 119, 228
novation (of designers) 11, 52, 53, 228

off-site overheads 61, 99, 100, 166
off-site payment bonds 156, 157, 158
omissions 76, 85, 93, 102
on-demand bond 156, 221, 230
on-site overheads 99, 166
open book (negotiation) 33, 51, 54, 56, 57, 228
open tender 24
ordering of materials (in claims) 101
outline design 11, 224
owner (of building) 6, 42, 121, 125–127, 129, 137, 144, 149, 231
ownership of materials 71, 117, 118, 121, 122, 140, 230

parametric (modelling of building work) 205
parent company guarantee 156
parity of tendering 42, 43, 228
partial possession 88, 89, 95, 153
partnering 1, 24, 29, 30, 31, 33, 35, 51, 228
partners (in partnering) 30
partners (business) 115
partnership(s) (business) 115, 116, 124
pay less notice 62, 66, 78, 92, 139, 201, 202, 228
pay when certified (terms in contracts) 195, 200, 203
pay when paid (terms in contracts) 200

performance bond 42, 155, 120, 121,
    155, 222
performance specification 10, 11, 24, 228
permitted purpose (in BIM) 211, 215
personal injury (insurance) 145, 146
photographs 84
planning permission 7, 11, 52
post-contract 13, 14, 46, 48, 50, 53, 55,
    133, 138, 179, 209, 221
post-tender checks 25, 41, 44, 228
postponement 91
Practical Completion 63, 66, 81, 87–91,
    94, 95, 153, 169, 171, 174, 180, 221
pre-tender checks 26, 27, 229
preambles 39, 222
preferred bidding 47, 48, 228
preliminaries (preliminary items) 39, 42,
    55, 56, 65, 67–69, 78, 82, 99, 101,
    143, 153, 178, 222, 225, 229, 230,
    232
price manipulation 43, 45
price ringing 26, 229
prime cost 19, 20, 22, 61, 225, 227
private finance initiative (PFI) 9, 10, 35,
    51, 60, 61, 231
pro-rata rates 76, 226, 229, 232
procurement Chapters 1 and 2, 2, 3, 60,
    74, 112, 119, 159, 231
professional (expert) client 147, 208, 226,
    229
professional indemnity insurance (PII) 3,
    13, 14, 152, 159, 198
project agreement (in BIM) 211
project manager 7, 34
property damage/damage to property
    (insurance) 145, 146, 147, 153, 158
provisional sum(s) 55, 78, 81, 82, 84,
    133, 147, 163, 218, 222, 225, 227,
    229, 232
public liability insurance 153
pure design and build 10, 11, 24, 60, 219

quantum meruit 98, 110, 111, 114

re-measurement 14
re-rating 77, 102, 229
rectification period 78, 89, 91, 177, 180
reductions 47, 67, 83, 222

relevant event(s) 92–95, 104, 105
relevant matter(s) 103, 104, 105
relevant omission 93
remoteness of damages 114
removal of work 170, 175
rescission 98, 110
reservation of title (building materials on/
    off site) 65, 121, 127, 140, 230
retention 63, 65, 66, 71, 91, 95, 123,
    140, 156–158, 174, 180, 230
retention bond 63, 66, 157, 158, 230
reverse auctioning 47, 230

Schedule 2 (variation/change) 77, 79, 80,
    83, 84, 94, 165
Schedule 2 (acceleration) 85, 96
Schedule 2 (named sub-contractor) 133,
    138
schedule(s) of rates 4, 16, 52, 53, 67, 79,
    80, 97, 99, 219, 224, 230
scheme for construction contracts 186,
    191, 198, 199, 202, 203
seasonal effect 100
sectional completion 88, 89
selective tendering 26, 27, 29, 30, 40, 41,
    43, 44
self-employed (sub-contractors) 132
serial tendering 28–30, 51
services engineer(s) 6, 129, 152, 209, 218
set-off 78, 92, 138–140, 180, 201, 226
shoe stepping (in taking action for latent
    defects) 130
single tender 31, 32, 35
site minutes 74
site overhead(s) 17, 20, 27, 61, 77, 99,
    100, 106, 118, 166, 229, 230
site survey(s) 38, 118
slump tests 181, 230
snagging (defects) 88, 131, 156, 171,
    177, 180, 230
soil investigations 39
soil reports 38
Special Project Vehicle(s) 60, 231
specialists (contractors) 7, 8, 20–23, 125,
    126–131, 133, 134, 137, 138, 141,
    143, 171, 177, 218, 219, 227, 231

specification (document) 4, 9, 12, 13, 38, 39, 53, 54, 80, 170, 173, 178, 210, 219, 224, 225
specified default 112
specified peril 95, 99, 114, 149, 151, 155, 227, 231
speculative pricing 43, 44, 46, 47
stage payment(s) 60, 61, 64, 67, 174, 199
standing list 25–27
statutory adjudication 186, 195, 196, 231
statutory insurance 153
statutory requirements 172, 178
strategic partnering 30, 33, 51
structural engineer(s) 6, 75, 209–211
subrogation 153, 154, 159, 231
substitution 79
suspend performance (under Construction Act) 63, 202

taking off 7, 14, 81, 205
target cost 17–19
technical check 45
tenant 3, 6, 42, 126, 128–130, 141, 149, 223, 231
tender type(s) 8, 23
tender adjudication 40
tender documents 24, 39, 40–42, 46, 52–54, 218, 225, 228, 232
tender drawings 39, 40
tender ratio 25, 27, 232
tender report 45
terrorism/terrorist acts 113, 151
tests 3, 63, 169, 170, 173, 175, 178, 179, 181
Third Party Rights Act 128, 129, 141
time-barred (action for redress for latent defects) 126, 127

tort 79, 128, 130, 141, 152, 198
trade contractor 134, 137
traditional contracting 13
traditional fluctuations 164, 166
transfer payment 22
transparency 22, 23, 51
two stage tendering 24, 33–35, 51, 57

undefined work 81, 232
uneconomic use of labour/working of labour 98, 100, 106, 223
uneconomic use of plant 101, 106
uneconomic ordering of materials 101, 106

valuation rules 76, 81, 82, 98, 232
valuing (valuation of) variations 14, 76, 79, 102, 163, 230, 232
variation ("Schedule 2") quotation 76, 77, 79, 80, 83
variation(s) 14, 43, 48, 50, 56, 60, 65, 67, 70, 73–84, 93, 94, 98, 101–103, 110, 115, 123, 124, 150, 152, 161–163, 165, 223, 230, 232
VAT 164, 165, 191

whole life costs 10, 51, 52, 232
winding up 116
work properly executed 62–64, 66, 67, 173, 223
work schedules 53, 67
working drawings 39, 232
Works Commencement Date 87
works contractor 20, 22, 133, 141
works insurance 3, 42, 63, 89, 117, 147, 148, 151
works orders 35
wrongful trading 116

Printed in the United States
By Bookmasters